Collins

ROYAL
OBSERVATORY
GREENWICH

2016 GUIDE
to the
NIGHT SKY

Storm Dunlop and Wil Tirion

Published by Collins
An imprint of HarperCollins Publishers
Westerhill Road
Bishopbriggs
Glasgow G64 2QT
www.harpercollins.co.uk

In association with
Royal Museums Greenwich, the group name for the National Maritime Museum,
Royal Observatory Greenwich, Queen's House and *Cutty Sark* 2015
www.rmg.co.uk

A catalogue record for this book is available from the British Library

ISBN 978-0-00-814131-8

10 9 8 7 6 5 4 3 2 1

Printed in China by RR Donnelley APS

If you would like to comment on any aspect of this book, please contact us at the above address or online.
e-mail: collinsmaps@harpercollins.co.uk
facebook.com/collinsmaps
@collinsmaps

Contents

Introduction

The aim of this Guide is to help people find their way around the night sky, by showing how the stars that are visible change from month to month and by including details of various events that occur throughout the year. The objects and events described may be observed with the naked eye, or nothing more complicated than a pair of binoculars.

The conditions for observing naturally vary over the course of the year. During the summer, twilight may persist throughout the night and make it difficult to see the faintest stars. There are three recognized stages of twilight: civil twilight, when the Sun is less than 6° below the horizon; nautical twilight, when the Sun is between 6° and 12° below the horizon; and astronomical twilight, when the Sun is between 12° and 18° below the horizon. Full darkness occurs only when the Sun is more than 18° below the horizon. During nautical twilight, only the very brightest stars are visible. During astronomical twilight, the faintest stars visible to the naked eye may be seen directly overhead, but are lost at lower altitudes. As the diagram shows, during the summer months full darkness never occurs at the latitude of London, and at Edinburgh

nautical twilight persists throughout the whole night, so at that latitude only the very brightest stars are visible.

Another factor that affects the visibility of objects is the amount of moonlight in the sky. At Full Moon, it may be very difficult to see some of the fainter stars and objects, and even when the Moon is at a smaller phase it may seriously interfere with visibility if it is near the stars or planets in which you are interested. A full lunar calendar is given for each month and may be used to see when nights are likely to be darkest and best for observation.

The celestial sphere

All the objects in the sky (including the Sun, Moon, and stars) appear to lie at some indeterminate distance on a large sphere, centred on the Earth. This *celestial sphere* has various reference points and features that are related to those of the Earth. If the Earth's rotational axis is extended, for example, it points to the North and South Celestial Poles, which are thus in line with the North and South Poles on Earth. Similarly, the *celestial equator* lies in the same plane as the Earth's equator, and divides the sky into northern and

The duration of twilight throughout the year at London and Edinburgh.

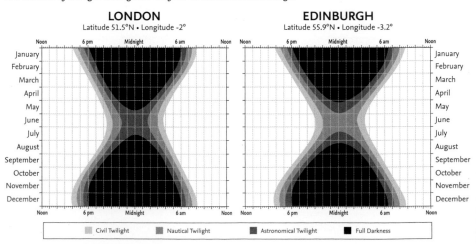

southern hemispheres. Because this Guide is written for use in Britain and Ireland, the area of the sky that it describes includes the whole of the northern celestial hemisphere and those portions of the southern that become visible at different times of the year. Stars in the far south, however, remain invisible throughout the year, and are not included.

It is useful to know some of the special terms for various parts of the sky. As seen by an observer, half of the celestial sphere is invisible, below the horizon. The point directly overhead is known as the **zenith**, and the (invisible) one below one's feet as the **nadir**. The line running from the north point on the horizon, up through the zenith and then down to the south point is the **meridian**. This is an important invisible line in the sky, because objects are highest in the sky, and thus easiest to see, when they cross the meridian in the south. Objects are said to **transit**, when they cross this line in the sky.

In this book, reference is frequently made in the text and in the diagrams to the standard compass points around the horizon. The position of any object in the sky may be described by its **altitude** (measured in degrees above the horizon), and its **azimuth** (measured in degrees from north 0°, through east 90°, south 180°, and west 270°). Experienced amateurs and professional astronomers also use another system of specifying locations on the celestial sphere, but that need not concern us here, where the simpler method will suffice.

The celestial sphere appears to rotate about an invisible axis, running between the North and South Celestial Poles. The location (i.e., the altitude) of the Celestial Poles depends entirely on the observer's position on Earth or, more specifically, their latitude. The charts in this book are produced for the latitude of 50°N, so the North Celestial Pole (NCP) is 50° above the northern horizon. The fact that the NCP is fixed relative to the horizon means that all the stars within 50° of the pole are always above the horizon and may, therefore, always be seen at night, regardless of the time of year. The northern circumpolar region is an ideal place to begin learning the sky, and ways to identify the circumpolar stars and constellations will be described shortly.

The ecliptic and the zodiac

Another important line on the celestial sphere is the Sun's apparent path against the background stars – in reality the result

Measuring altitude and azimuth on the celestial sphere.

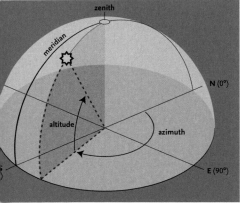

The altitude of the North Celestial Pole equals the observer's latitude.

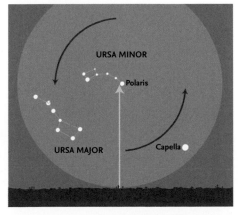

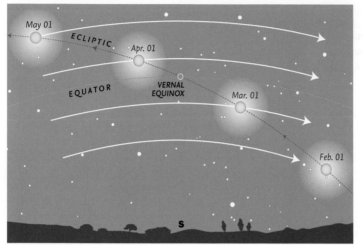

The Sun crossing the celestial equator in spring.

of the Earth's orbit around the Sun. This is known as the *ecliptic*. The point where the Sun, apparently moving along the ecliptic, crosses the celestial equator from south to north is known as the vernal (or spring) equinox, which occurs on March 20. At this time (and at the autumnal equinox, on September 22 or 23, when the Sun crosses the celestial equator from north to south) day and night are almost exactly equal in length. (There is a slight difference, but that need not concern us here.) The vernal equinox is currently located in the constellation of Pisces, and is important in astronomy because it defines the zero point for a system of celestial coordinates, which is, however, not used in this Guide.

The Moon and planets are to be found in a band of sky that extends 8° on either side of the ecliptic. This is because the orbits of the Moon and planets are inclined at various angles to the ecliptic (i.e., to the plane of the Earth's orbit). This band of sky is known as the Zodiac and, when originally devised, consisted of twelve **constellations**, all of which were considered to be exactly 30° wide. When the constellation boundaries were formally established by the International Astronomical Union in 1930, the exact extent of most constellations was altered and, nowadays, the ecliptic passes through thirteen constellations. Because of the boundary changes, the Moon and planets may actually pass through several other constellations that are adjacent to the original twelve.

The constellations

Since ancient times, the celestial sphere has been divided into various constellations, most dating back to antiquity and usually associated with certain myths or legendary people and animals. Nowadays, the boundaries of the constellations have been fixed by international agreement and their names (in Latin) are largely derived from Greek or Roman originals. Some of the names of the most prominent stars are of Greek or Roman origin, but many are derived from Arabic names. Many bright stars have no individual names and, for many years, stars were identified by terms such as 'the star in Hercules' right foot'. A more sensible scheme was introduced by the German astronomer Johannes Bayer in the early seventeenth century. Following his scheme – which is still used today – most of the brightest stars are identified by a Greek letter followed by the genitive form of the constellation's Latin name. An example is the Pole Star, also known as Polaris and α Ursae Minoris (abbreviated α UMi). The Greek alphabet is shown on page 93 with a list of all the constellations that may be seen from latitude 50°N, together with abbreviations, their genitive forms and English names. Other naming schemes exist for fainter stars, but are not used in this book.

Asterisms

Apart from the constellations (88 of which cover the whole sky), certain groups of stars, which may form a part of a larger constellation or cross several constellations, are readily recognizable and have been given individual names. These groups are known as *asterisms*, and the most famous (and well-known) is the 'Plough', the common name for the seven brightest stars in the constellation of Ursa Major, the Great Bear. The names and details of some asterisms mentioned in this book are given in the list on page 94.

Magnitudes

The brightness of a star, planet or other body is frequently given in magnitudes (mag.). This is a mathematically defined scale where larger numbers indicate a fainter object. The scale extends beyond the zero point to negative numbers for very bright objects. (Sirius, the brightest star in the sky is mag. -1.4.) Most observers are able to see stars down to about mag. 6, under very clear skies.

The Moon

As it gradually passes across the sky from west to east in its orbit around the Earth, the Moon moves by approximately its diameter (about half a degree) in an hour. Normally, in its orbit around the Earth, the Moon passes above or below the direct line between Earth and Sun (at New Moon) or outside the area obscured by the Earth's shadow (at Full Moon).

Occasionally, however, the three bodies are more-or-less perfectly aligned to give an eclipse: a solar eclipse at New Moon or a lunar eclipse at Full Moon. Depending on the exact circumstances, a solar eclipse may be merely partial (when the Moon does not cover the whole of the Sun's disk); annular (when the Moon is too far from Earth in its orbit to appear large enough to hide the whole of the Sun); or total. Total and annular eclipses are visible from very restricted areas of the Earth, but partial eclipses are normally visible over a wider area.

Somewhat similarly, at a lunar eclipse, the Moon may pass through the outer zone of the Earth's shadow, the **penumbra** (in a penumbral eclipse, which is not generally perceptible to the naked eye), so that just part of the Moon is within the darkest part of the Earth's shadow, the **umbra** (in a partial eclipse); or completely within the umbra (in a total eclipse). Unlike solar eclipses, lunar eclipses are visible from large areas of the Earth.

Occasionally, as it moves across the sky, the Moon passes between the Earth and individual planets or distant stars giving rise to an **occultation**. As with solar eclipses, such occultations are visible from restricted areas of the world.

The Planets

Because the planets are always moving against the background stars, they are treated in some detail in the monthly pages and information is given when they are close to other planets, the Moon or any of five bright stars that lie near the ecliptic. Such events are known as **appulses** or, more frequently, as **conjunctions**. (There are technical differences in the way these terms are defined – and should be used – in astronomy, but these need not concern us here.) The positions of the planets are shown for every month on a special chart of the ecliptic.

The term conjunction is also used when a planet is either directly behind or in front of

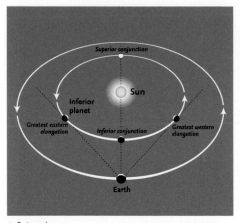

Inferior planet.

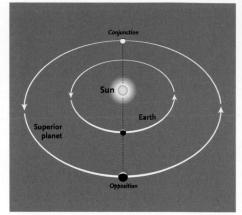

Superior planet.

the Sun, as seen from Earth. (Under normal circumstances it will then be invisible.) The conditions of most favourable visibility depend on whether the planet is one of the two known as **inferior planets** (Mercury and Venus) or one of the three **superior planets** (Mars, Jupiter and Saturn) that are covered in detail. (Some details of the fainter superior planets, Uranus and Neptune, are included in this Guide, and special charts for both are given on page 18.)

The inferior planets are most readily seen at eastern or western **elongation**, when their angular distance from the Sun is greatest. For superior planets, they are best seen at **opposition**, when they are directly opposite the Sun in the sky, and cross the meridian at local midnight.

It is often useful to be able to estimate angles on the sky, and approximate values may be obtained by holding one hand at arm's length. The various angles are shown in the diagram, together with the separations of the various stars in the Plough.

Meteors

At some time or other, nearly everyone has seen a **meteor** – a 'shooting star' – as it flashed across the sky. The particles that cause meteors – known technically as 'meteoroids' – range in size from that of a grain of sand (or even smaller), to the size of a pea. On any night of the year there are occasional meteors, known as **sporadics**, that may travel in any direction. These occur at a rate that is normally between three and eight in an hour. Far more important, however, are **meteor showers**, which occur at fixed periods of the year, when the Earth encounters a trail of particles left behind by a comet or, very occasionally, by a minor planet (asteroid). Meteors always appear to diverge from a single point on the sky, known as the **radiant**, and the radiants of major showers are shown on the charts. Meteors that come from a circular area 8° in diameter around the radiant are classed as belonging to the particular shower. All others that do not come from that area are sporadics (or, occasionally from another shower that is active at the same time). A list of the major meteor showers is given on page 17.

Although the positions of the various shower radiants are shown on the charts, looking directly at the radiant is not the most effective way of seeing meteors. They are most likely to be noticed if one is looking about 40–45° away from the radiant position. (This is approximately two hand-spans as shown in the diagram for measuring angles.)

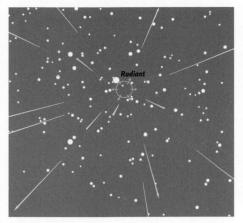

Meteor shower.

Measuring angles in the sky.

Other objects

Certain other objects may be seen with the naked eye under good conditions. Some were given names in antiquity – Praesepe is one example – but many are known by what are called 'Messier numbers', the numbers in a catalogue of nebulous objects compiled by Charles Messier in the late eighteenth century. Some, such as the Andromeda Galaxy, M31, and the Orion Nebula, M42, may be seen by the naked eye, but all those given in the list will benefit from the use of binoculars. Apart from galaxies, such as M31, which

contain thousands of millions of stars, there are also two types of cluster: open clusters, such as M45, the Pleiades, which may consist of a few dozen to some hundreds of stars; and globular clusters, such as M13 in Hercules, which are spherical concentrations of many thousands of stars. One or two gaseous nebulae, consisting of gas illuminated by stars within them, are also visible. The Orion Nebula, M42, is one, and is illuminated by the group of four stars, known as the Trapezium, which may be seen within it by using a good pair of binoculars.

Some interesting objects.

Messier / NGC	Name	Type	Constellation	Maps (months)
—	Hyades	open cluster	Taurus	Sep. – Apr.
—	Double Cluster	open cluster	Perseus	All year
—	Melotte 111 (Coma Cluster)	open cluster	Coma Berenices	Jan. – Aug.
M3	—	globular cluster	Canes Venatici	Jan. – Sep.
M4	—	globular cluster	Scorpius	May – Aug.
M8	Lagoon Nebula	gaseous nebula	Sagittarius	Jun. – Sep.
M11	Wild Duck Cluster	open cluster	Scutum	May – Oct.
M13	Hercules Cluster	globular cluster	Hercules	Feb. – Nov.
M15	—	globular cluster	Pegasus	Jun. – Dec.
M22	—	globular cluster	Sagittarius	Jun. – Sep.
M27	Dumbbell Nebula	planetary nebula	Vulpecula	May – Dec.
M31	Andromeda Galaxy	galaxy	Andromeda	All year
M35	—	open cluster	Gemini	Oct. – May
M42	Orion Nebula	gaseous nebula	Orion	Nov. – Mar.
M44	Praesepe	open cluster	Cancer	Nov. – Jun.
M45	Pleiades	open cluster	Taurus	Aug. – Apr.
M57	Ring Nebula	planetary nebula	Lyra	Apr. – Dec.
M67	—	open cluster	Cancer	Dec. – May
NGC 752	—	open cluster	Andromeda	Jul. – Mar.
NGC 3242	Ghost of Jupiter	planetary nebula	Hydra	Feb. – May

The Northern Circumpolar Constellations

The northern circumpolar stars are the key to starting to identify the constellations. For anyone in the northern hemisphere they are visible at any time of the year, and nearly everyone is familiar with the seven stars of the Plough – known as the Big Dipper in North America – an asterism that forms part of the large constellation of Ursa Major (the Great Bear).

Ursa Major

Because of the movement of the stars caused by the passage of the seasons, Ursa Major lies in different parts of the evening sky at different periods of the year. The diagram below shows its position for the four main seasons. The seven stars of the Plough remain visible throughout the year anywhere north of latitude 40°N. Even at the latitude (50°N) for which the charts in this book are drawn, many of the stars in the southern portion of the constellation of Ursa Major are hidden below the horizon for part of the year or (particularly in late summer) cannot be seen late in the night.

Polaris and Ursa Minor

The two stars **Dubhe** and **Merak** (α and β

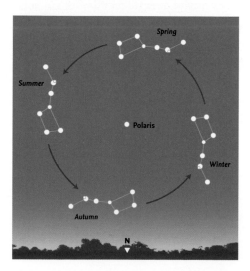

Ursae Majoris, respectively), farthest from the 'tail' are known as the 'Pointers'. A line from Merak to Dubhe, extended about five times their separation, leads to the Pole Star, **Polaris**, or α Ursae Minoris. All the stars in the northern sky appear to rotate around it. There are five main stars in the constellation of **Ursa Minor**, and the two farthest from the Pole, **Kochab** and **Pherkad**, (β and γ Ursae Minoris, respectively) are known as 'The Guards'.

Cassiopeia

On the opposite of the North Pole from Ursa Major lies **Cassiopeia**. It is highly distinctive, appearing as five stars forming a letter 'W' or 'M' depending on its orientation. Provided the sky is reasonably clear of clouds, you will nearly always be able to see either Ursa Major or Cassiopeia, and thus be able to orientate yourself on the sky.

To find Cassiopeia, start with **Alioth** (ε Ursae Majoris), the first star in the tail of the Great Bear. A line from this star extended through Polaris points directly towards **Cih** (γ Cassiopeiae), the central star of the five.

Cepheus

Although the constellation of **Cepheus** is fully circumpolar, it is not nearly as well-known as Ursa Major, Ursa Minor or Cassiopeia, partly because its stars are fainter. Its shape is rather like the gable end of a house. The line from the Pointers through Polaris, if extended, leads to **Errai** (γ Cephei) at the 'top' of the 'gable'. The brightest star, **Alderamin** (α Cephei) lies in the Milky-Way region, at the 'bottom right-hand corner' of the figure.

Draco

The constellation of **Draco** consists of a quadrilateral of stars, known as the 'Head of Draco' (and also the 'Lozenge'), and a long chain of stars forming the neck and body of the dragon. To find the Head of Draco, locate the two stars Phad and Megrez (γ and δ Ursae Majoris) in the Plough, opposite the

The stars and constellations inside the circle are always above the horizon, seen from our latitude.

Pointers. Extend a line from Phad through Megrez by about eight times their separation, right across the sky below the Guards in Ursa Minor, ending at **Grumium** (ξ Draconis) at one corner of the quadrilateral. The brightest star, **Etamin** (γ Draconis) lies farther to the south.

From the head of Draco, the constellation first runs northeast to **Altais** (δ Draconis) and ε Draconis, then doubles back southwards before winding its way through **Thuban** (α Draconis) before ending at Giausar (λ Draconis) between the Pointers and Polaris.

The Winter Constellations

The winter sky is dominated by several bright stars and distinctive constellations. The most conspicuous constellation is **Orion**, the main body of which has an hour-glass shape. It straddles the celestial equator and is thus visible from anywhere in the world. The three stars that form the 'Belt' of Orion point down towards the southeast and to **Sirius** (α Canis Majoris), the brightest star in the sky. **Mintaka** (δ Orionis), the star at the northeastern end of the Belt, farthest from Sirius, actually lies just slightly south of the celestial equator.

A line from **Bellatrix** (γ Orionis) at the 'top right-hand corner' of Orion, through **Aldebaran** (α Tauri), past the 'V' of the Hyades cluster, points to the distinctive cluster of bright blue stars known as the **Pleiades**, or the 'Seven Sisters'. Aldebaran is one of the five bright stars that may sometimes be occulted (hidden) by the Moon. Another line from Bellatrix, through

Betelgeuse (α Orionis), if carried right across the sky, points to the constellation of **Leo**, a prominent constellation in the spring sky.

Six bright stars in six different constellations: **Capella** (α Aurigae), **Aldebaran** (α Tauri), **Rigel** (β Orionis), **Sirius** (α Canis Majoris), **Procyon** (α Canis Minoris) and **Pollux** (β Gemini) form what is sometimes known as the 'Winter Hexagon'. Pollux is accompanied to the northwest by the slightly fainter star of **Castor** (α Gemini), the second 'Twin'.

In a counterpart to the famous 'Summer Triangle', an almost perfect equilateral triangle, the 'Winter Triangle', is formed by Betelgeuse (α Orionis), Sirius (α Canis Majoris) and Procyon (α Canis Minoris).

Several of the stars in this region of the sky show distinctive tints: Betelgeuse (α Orionis) is reddish, Aldebaran (α Tauri) is orange, and Rigel (β Orionis) is blue-white.

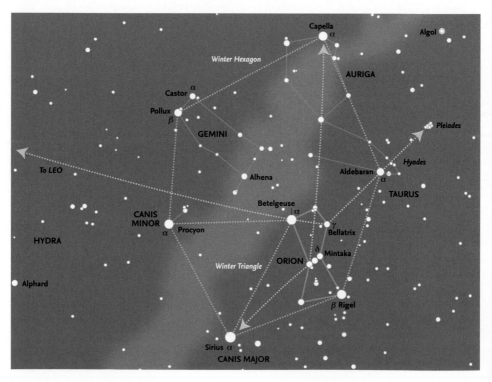

The Spring Constellations

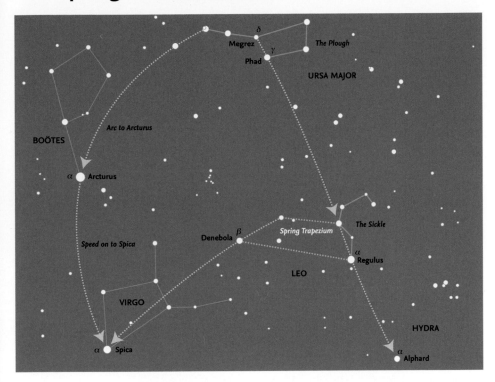

The most prominent constellation in the spring sky is the zodiacal constellation of **Leo**, and its brightest star, **Regulus** (α Leonis), which may be found by extending a line from Megrez and Phad (δ and γ Ursae Majoris, respectively) – the two stars on the opposite side of the bowl of the Plough from the Pointers – down to the southeast. Regulus forms the 'dot' of the 'backward question mark' known as 'the Sickle'. Regulus, like Aldebaran in Taurus is one of the bright stars that lie close to the ecliptic, and which are occasionally occulted by the Moon. The same line from Ursa Major to Regulus, if continued, leads to **Alphard** (α Hydrae), the brightest star in **Hydra**, the largest of the 88 constellations.

The shape formed by the body of Leo is sometimes known as the 'Spring Trapezium'. At the other end of the constellation from Regulus is **Denebola** (β Leonis), and the line

forming the back of the constellation through Denebola, points to the bright star **Spica** (α Virginis) in the constellation of **Virgo**. A saying that helps to locate Spica is well-known to astronomers: 'Arc to Arcturus and then speed on to Spica.' This suggests following the arc of the tail of Ursa Major to Arcturus and then on to Spica. **Arcturus** (α Boötis) is actually the brightest star in the northern hemisphere of the sky. (Although other stars, such as Sirius, are brighter, they are all in the southern hemisphere.) Overall, the constellation of **Boötes** is sometimes described as 'kite-shaped' or 'shaped like the letter P'.

Although Spica is the brightest star in the zodiacal constellation of Virgo, the rest of the constellation is not particularly distinct, consisting of a rough quadrilateral of moderately bright stars and some fainter lines of stars extending outwards.

The Summer Constellations

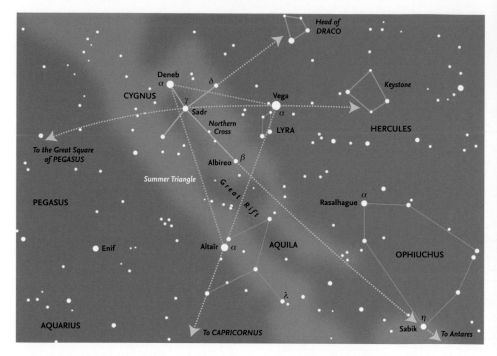

On summer nights, the three bright stars **Deneb** (α Cygni), **Vega** (α Lyrae) and **Altair** (α Aquilae) form the striking 'Summer Triangle'. The constellations of **Cygnus** (the Swan) and **Aquila** (the Eagle) represent birds 'flying' down the length of the Milky Way. This part of the Milky Way contains the **Great Rift**, an elongated dark region, where the light from distant stars is obscured by intervening dust. The dark Rift is clearly visible even to the naked eye.

The most prominent stars of Cygnus are sometimes known as the 'Northern Cross' (as a counterpart to the 'Southern Cross' – the constellation of Crux – in the southern hemisphere). The central line of Cygnus through **Albireo** (β Cygni), extended well to the southwest, points to **Sabik** (η Ophiuchi) in the large, sprawling constellation of Ophiuchus (the Serpent Bearer) and beyond to **Antares** (α Scorpii) in the constellation of **Scorpius**. Like Cepheus, the shape of Ophiuchus somewhat resembles the gable-end of a house, and the

brightest star **Rasalhague** (α Ophiuchi) is at the 'apex' of the 'gable'.

A line from the central star of Cygnus, **Sadr** (γ Cygni) through δ Cygni, in the northwestern 'wing' points towards the Head of Draco, and is another way of locating that part of the constellation. Another line from Sadr to Vega indicates the central portion, 'The Keystone', of the constellation of **Hercules**. An arc through the same stars, in the opposite direction, points towards the constellation of **Pegasus**, and more specifically to the 'Great Square of Pegasus'.

Aquila is less conspicuous than Cygnus and consists of a diamond shape of stars, representing the body and wings of the eagle, together with a rather faint star, λ Aquilae, marking the 'head'. **Lyra** (the Lyre) mainly consists of Vega (α Lyrae) and a small quadrilateral of stars to its southeast. Continuation of a line from Vega through Altair indicates the zodiacal constellation of **Capricornus**.

14

The Autumn Constellations

During the autumn season, the most striking feature is the 'Great Square of Pegasus', an almost perfect rectangle on the sky, forming the main body of the constellation of **Pegasus**. However, the star at the northeastern corner, **Alpheratz**, is actually α Andromedae, and part of the adjacent constellation of **Andromeda**. A line from **Scheat** (β Pegasi) at the northeastern corner of the Square, through **Matar** (η Pegasi), points in the general direction of Cygnus. A crooked line of stars leads from **Markab** (α Pegasi) through **Homam** (ζ Pegasi) and **Biham** (θ Pegasi) to **Enif** (ε Pegasi). A line from Markab through the last star in the Square, **Algenib** (γ Pegasi) points in the general direction of the five stars, including **Menkar** (α Ceti) that form the 'tail' of the constellation of **Cetus** (the Whale). A ring of seven stars lying below the southern side of the Great Square is known as 'the Circlet', part of the constellation of **Pisces** (the Fishes).

Extending the line of the western side of the Great Square towards the south leads to the isolated bright star, **Fomalhaut** (α Piscis Austrini), in the Southern Fish. Following the line of the eastern side of the Great Square towards the north leads to Cassiopeia while, in the other direction, it points towards **Deneb Kaitos** (β Ceti), which is actually the brightest star in Cetus.

Three bright stars leading northeast from Alpheratz form the main body of the constellation of **Andromeda**. Continuation of that line leads towards the constellation of **Perseus** and **Mirphak** (α Persei). Running southwards from Mirphak is a chain of stars, one of which is the famous variable star **Algol** (β Persei). Farther east, an arc of stars leads to the prominent cluster of the **Pleiades**, in the constellation of **Taurus.**

Between Andromeda and Cetus lie the two small constellations of **Triangulum** (the Triangle) and **Aries** (the Ram).

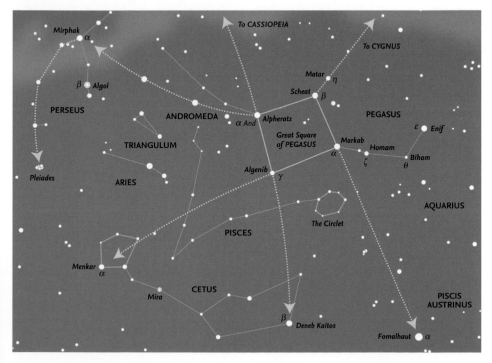

Introduction to the Month-by-Month Guide

The monthly charts

The pages devoted to each month contain a pair of charts showing the appearance of the night sky, looking north and looking south. The charts (as with all the charts in this book) are drawn for the latitude of 50°N, so observers farther north will see slightly more of the sky on the northern horizon, and slightly less on the southern. These areas are, of course, those most likely to be affected by poor observing conditions caused by haze, mist or smoke. In addition, stars close to the horizon are always dimmed by atmospheric absorption, so sometimes the faintest stars marked on the charts may not be visible.

The three times shown for each chart require a little explanation. The charts are drawn to show the appearance at 23:00 GMT for the 1st of each month. The same appearance will apply an hour earlier (22:00 GMT) on the 15th, and yet another hour earlier (21:00 GMT) at the end of the month (shown as the 1st of the following month). GMT is identical to the Universal Time (UT), used by astronomers around the world. In Europe, Summer Time is introduced in March, so the March charts apply to 23:00 GMT on March 1, 22:00 GMT on March 15, but 22:00 BST (British Summer Time) on April 1. The change back from Summer Time (in Europe) occurs in October, so the charts for that month apply to 00:00 BST for October 1, 23:00 BST for October 15, and 21:00 GMT for November 1.

The charts may be used for earlier or later times during the night. To observe two hours earlier, use the charts for the preceding month; for two hours later, the charts for the next month.

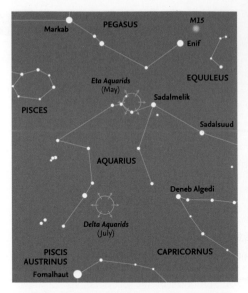

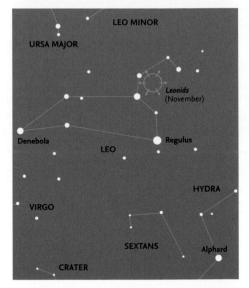

Meteors

Details of specific meteor showers are given in the months when they come to maximum, regardless of whether they begin or end in other months. Note that not all the respective radiants are marked on the charts for that particular month, because the radiants may be below the horizon, or lie in constellations that are not readily visible during the month of maximum. For this reason, special charts for the Eta and Delta Aquarids (May and July, respectively) and the Leonids (November)

Shower	Dates of activity 2016	Date of maximum 2016	Possible hourly rate
Quadrantids	January 1–10	January 3–4	120
April Lyrids	April 16–25	April 22–23	18
Eta Aquarids	April 19 to May 26	May 6–7	55
Alpha Capricornids	July 11 to August 10	July 27–28	5
Perseids	July 13 to August 26	August 12–13	100
Delta Aquarids	July 21 to August 23	July 28–29	< 20
Alpha Aurigids	August to October	August 28 & September 15	10
Southern Taurids	September 7 to November 19	October 23–24	< 5
Northern Taurids	October 19 to December 10	November 11–12	< 5
Orionids	October 4 to November 14	October 21–22	25
Leonids	November 5–30	November 17–18	< 15
Geminids	December 4–16	December 13–14	100+
Ursids	December 17–23	December 21–22	< 10

are given here. As explained earlier, however, meteors from such showers may still be seen, because the most effective region for seeing meteors is some 40–45° away from the radiant, and that area of sky may well be above the horizon. A table of the best meteor showers visible during the year is also given here.

The photographs

As an aid to identification – especially as some people find it difficult to relate charts to the actual stars they see in the sky – one or more photographs of constellations visible in certain specific months are included. It should be noted, however, that because of the limitations of the photographic and printing processes, and the differences between the sensitivity of different individuals to faint starlight (especially in their ability to detect different colours), and the degree to which they have become adapted to the dark, the apparent brightness of stars in the photographs will not necessarily precisely match that seen by any one observer.

The Moon calendar

The Moon calendar is largely self-explanatory. It shows the phase of the Moon for every day of the month, with the exact times (in Universal Time) of New Moon, First Quarter, Full Moon, and Last Quarter. Because the times are calculated from the Moon's actual orbital parameters, some of the times shown will, naturally, fall during daylight, but any difference is too small to affect the appearance of the Moon on that date.

The Moon

The section on the Moon includes details of any lunar or solar eclipses that may occur during the month (visible from anywhere on Earth). Similar information is given about any important occultations. Mainly, however, this section summarizes when the Moon passes close to planets or the five prominent stars close to the ecliptic. The dates when the Moon is closest to the Earth (at *perigee*) and farthest from it (at *apogee*) are shown in the monthly calendars, and only mentioned here when they are particularly significant, such as the nearest and farthest during the year.

The Planets and minor planets

Brief details are given of the location, movement and brightness of the planets from Mercury to Saturn throughout the month. None of the planets can, of course, be seen when they are close to the Sun, so such periods are generally noted. All of the planets may sometimes lie on the opposite side of the Sun to the Earth (at superior conjunction), but in the case of the inferior planets, Mercury and Venus, they may also pass between the

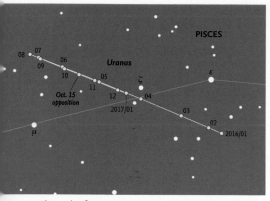

The path of Uranus in 2016. Uranus comes to opposition on October 15.

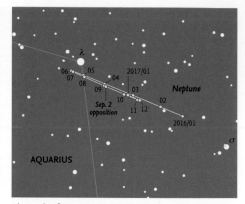

The path of Neptune in 2016. Neptune comes to opposition on September 2.

Earth and the Sun (at inferior conjunction) and are normally invisible for a longer or shorter period of time. Those two planets are normally easiest to see around either eastern or western elongation, in the evening or morning sky, respectively. Not every elongation is favourable, so although every elongation is listed, only those where observing conditions are favourable are shown in the individual diagrams of events.

The dates at which the superior planets reverse their motion (from direct motion to retrograde, and retrograde to direct) and of opposition (when a planet generally reaches its maximum brightness) are given. Some planets, especially distant Saturn, may spend most or all of the year in a single constellation. Jupiter and Saturn are normally easiest to see around opposition which occurs every year. Mars, by contrast, moves relatively rapidly against the background stars and in some years never comes to opposition.

Uranus is not always included in the monthly details because it is generally at the limit of naked-eye visibility (magnitude 5.7–5.9), although bright enough to be visible in binoculars, or even with the naked eye under exceptionally dark skies. Its path in 2016 is shown on the special chart above. It comes to

opposition on October 15 in the constellation of Pisces. Unfortunately, Full Moon is one day later, so the planet, at magnitude 5.7 will be difficult to detect. About a week on either side of that date, it should be visible, when its change in position relative to the background stars will become noticeable.

Similar considerations apply to Neptune, although this is always fainter (magnitude 7.8–8.0 in 2016), but still visible in most binoculars. It reaches opposition at mag. 7.8 on September 2 in Aquarius. The Moon is at Last Quarter on the same day, but should not cause too much interference. Again, Neptune's path in 2016 and its position at opposition are shown in the chart above.

Charts for the three brightest minor planets that come to opposition in 2016 are shown in the relevant month: Juno (mag. 10.0) on April 27, Pallas (mag. 9.2) on August 20, and Ceres (mag. 7.8) on October 21.

The ecliptic charts

Although the ecliptic charts are primarily designed to show the positions and motions of the major planets, they also show the motion of the Sun during the month. The light-tinted area shows the area of the sky that is invisible during daylight, but the darker area gives an indication of which constellations are likely to be visible at some time of the night. The closer a planet is to the border between dark and light, the more difficult it will be to see in the twilight.

The monthly calendar

For each month, a calendar shows details of significant events, including when planets are close to one another in the sky, close to the Moon, or close to any one of five bright stars that are spaced along the ecliptic. The times shown are given in Universal Time (UT), always used by astronomers throughout the year, and which is identical to Greenwich Mean Time (GMT). So during the summer months, they do not show Summer Time, which will always be one hour later than the time shown.

The diagrams of interesting events

Each month, a number of diagrams show the appearance of the sky when certain events take place. However, the exact positions of celestial objects and their separations greatly depend on the observer's position on Earth. When the Moon is one of the objects involved, because it is relatively close to Earth, there may be very significant changes from one location to another. Close approaches between planets or between a planet and a star are less affected by changes of location, which may thus be ignored.

The diagrams showing the appearance of the sky are drawn for the latitude of London, so will be approximately correct for most of Britain and Europe. However, for an observer farther north (say Edinburgh), a planet or star listed as being north of the Moon will appear even farther north, whereas one south of the Moon will appear closer to it – or may even be hidden (occulted) by it. For an observer farther south than London, there will be corresponding changes in the opposite direction: for a star or planet south of the Moon the separation will increase, and for one north of the Moon the separation will decrease. For example, on 20 January 2016 during the occultation of Aldebaran (pages 21 and 25), the apparent path of the star appears farther north seen from Edinburgh than from London.

Ideally, details should be calculated for each individual observer, but this is obviously impractical. In fact, positions and separations are actually calculated for a theoretical observer located at the centre of the Earth.

So the details given regarding the positions of the various bodies should be used as a guide to their location. A similar situation arises with the times that are shown. These are calculated according to certain technical criteria, which need not concern us here. However, they do not necessarily indicate the exact time when two bodies are closest together. Similarly, dates and times are given, even if they fall in daylight, when the objects are likely to be completely invisible. However, such times do give an indication that the objects concerned will be in the same general area of the sky during both the preceding, and the following nights.

Key to the symbols used on the monthy star maps.

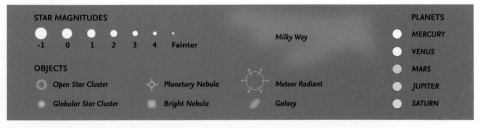

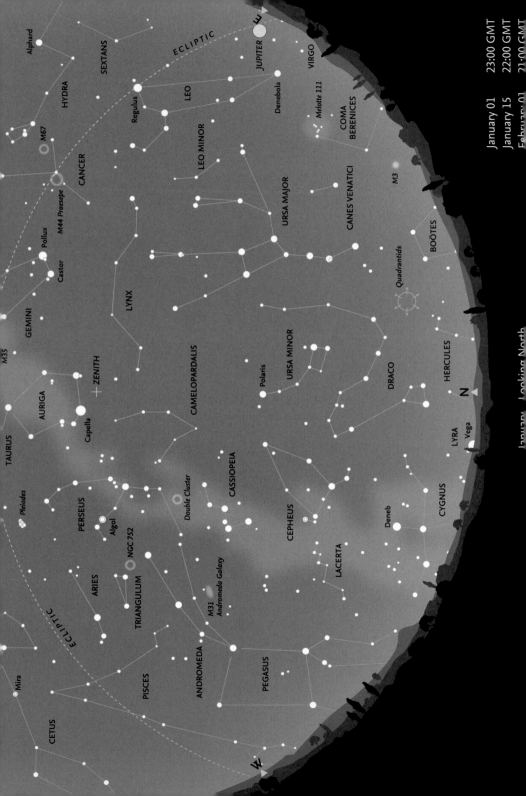

January, Looking North

January 01   23:00 GMT
January 15   22:00 GMT
February 01   21:00 GMT

January – Looking North

Most of the important circumpolar constellations are easy to see in the northern sky at this time of year. *Ursa Major* stands more-or-less vertically above the horizon in the northeast, with the zodiacal constellation of *Leo* rising in the east. To the north, the stars of *Ursa Minor* lie below *Polaris* (the Pole Star). The head of *Draco* is low on the northern horizon, but may be difficult to see unless observing conditions are good. Both *Cepheus* and *Cassiopeia* are readily visible in the northwest, and even the faint constellation of *Camelopardalis* is high enough in the sky for it to be easily visible. *Auriga*, with bright *Capella* is high overhead, near the zenith, while *Perseus* and *Andromeda* are visible farther down towards the west, where the Great Square of *Pegasus* is approaching the horizon.

Meteors

One of the strongest and most consistent meteor showers occurs in January: the Quadrantids, which are visible January 1–10, with maximum on January 3–4. They are brilliant, bluish and yellowish-white meteors and at maximum may even reach a rate of 120 meteors per hour. At maximum, the Moon is a waning crescent, just past Last Quarter, so there should not be too much interference from moonlight. The parent object is minor planet 2003 EH.

The shower is named after the former constellation *Quadrans Muralis* (the Mural Quadrant), an early form of astronomical instrument. The Quadrantid meteor radiant is now within the northernmost part of *Boötes*, roughly half-way between θ Boötis and τ Herculis.

Occultations

Of the bright stars near the ecliptic that may be occulted by the Moon (Aldebaran, Antares, Pollux, Regulus, and Spica), Aldebaran is occulted 14 times in 2016, just two of which are (theoretically) visible from Britain, and Regulus once, visible only from the far southwest of Australia.

The occultation of Aldebaran on January 20 is extremely low, and reappearance occurs after the Moon and star have set. On December 13 there is a graze of one component – Aldebaran is double – but the Moon is less than one day from Full, so observations are extremely difficult, even for experienced observers.

M31, the great Andromeda Galaxy, is about 220,000 light-years across, but has recently been found to be surrounded by an otherwise invisible halo of hot (10,000 to 100,000 degrees) ionized hydrogen at least 5 times that diameter, reaching half-way to our own Galaxy.

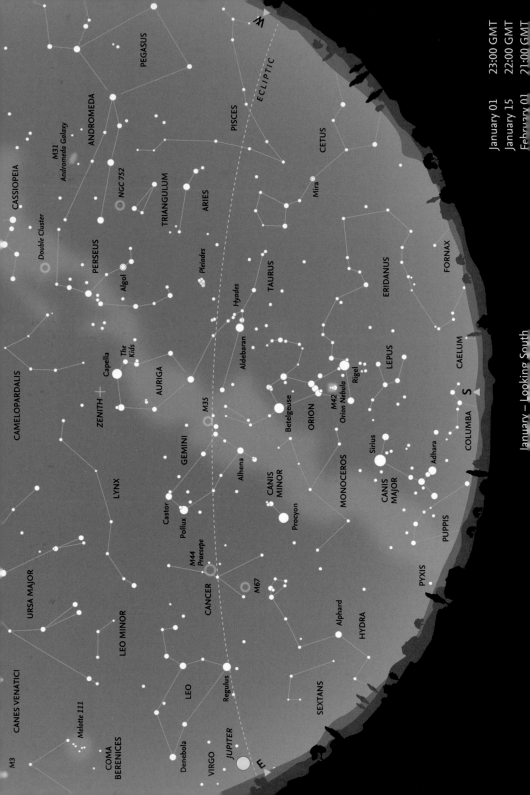

January 01   23:00 GMT
January 15   22:00 GMT
February 01   21:00 GMT

January – Looking South

January – Looking South

At this time of year the southern sky is dominated by **Orion.** This is the most prominent constellation during the winter months, when it is visible at some time during the night. (A photograph of Orion appears on page 87.) It has a highly distinctive shape, with a line of three stars that form the 'Belt'. To most observers, the bright star at the northwestern corner of the constellation, **Betelgeuse** (α Orionis), shows a reddish tinge, in contrast to the brilliant bluish-white colour of the bright star at the southwestern corner, **Rigel** (β Orionis). The three stars of the belt lie directly south of the celestial equator. A vertical line of three 'stars' forms the 'Sword' that hangs to the south of the Belt. With good viewing conditions, the central 'star' appears as a hazy spot, even to the naked eye. This is actually the Orion Nebula. Binoculars will reveal the four stars of the Trapezium, which illuminate the nebula.

The line of Orion's Belt points up to the northwest towards **Taurus** (the Bull) and below orange-tinted **Aldebaran** (α Tauri). Close to Aldebaran, there is a conspicuous 'V' of stars, pointing down to the southwest, called the **Hyades** cluster. (Despite appearances, Aldebaran

is not part of the cluster.) Farther along, the same line from Orion passes below a bright cluster of stars, the **Pleiades,** or Seven Sisters. Even the smallest pair of binoculars reveals this cluster to be a beautiful group of bluish-white stars. The two most conspicuous of the other stars in Taurus lie directly above Orion, and form an elongated triangle with Aldebaran. The northernmost, β Tauri, was once considered to be part of the constellation of Auriga.

Also above Orion at this time of year is the constellation of **Auriga** (the Charioteer), with brilliant **Capella** (α Aurigae), close to the zenith, directly overhead. Slightly to the west of Capella lies a small triangle of fainter stars, known as 'The Kids'. (Ancient mythological representations of Auriga show him carrying two young goats.) Together with that northernmost bright star in Taurus, β Tauri, the body of Auriga forms a large pentagon on the sky, with The Kids lying on the western side.

The Moon's phases for January

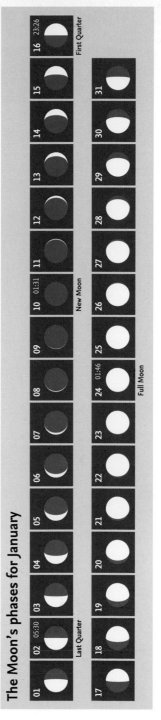

January – Moon and Planets

The Earth

The Earth reaches perihelion (the closest point to the Sun in its annual orbit) on 2 January 2016, at 22:49 Universal Time. Its distance is then 0.9833 AU (147,100,175 km).

The Moon

On January 2, the Last-Quarter Moon is in **Virgo**. By January 4 it is close to **Mars** and on January 5–6 passes close to **Venus** and **Saturn**, above **Antares** in the morning sky. On January 20 there is an occultation of Aldebaran, but this is very low in the western sky and only the disappearance is visible before the Moon and star set.

The Planets

Mercury is too close to the Sun to be seen, passing between the Sun and Earth at inferior conjunction on January 14. **Venus** is initially bright (magnitude -4.0) on the very border of **Libra** and **Scorpius**, but rapidly approaches the Sun and becomes invisible in the morning sky. **Mars** is also a morning object, initially in **Virgo** at magnitude 1.3, but brightening slightly to magnitude 0.9 as it enters **Libra**. Jupiter is retrograding slowly in **Leo** at magnitude -2.2 to -2.4 over the month. **Saturn** (magnitude 0.5), low in the morning sky, moves very slowly eastwards in **Ophiuchus**. **Uranus** (magnitude 5.8–5.9) is in **Pisces**, where it remains throughout the year. Similarly, Neptune lies in **Aquarius** and is magnitude 7.9–8.0 in January.

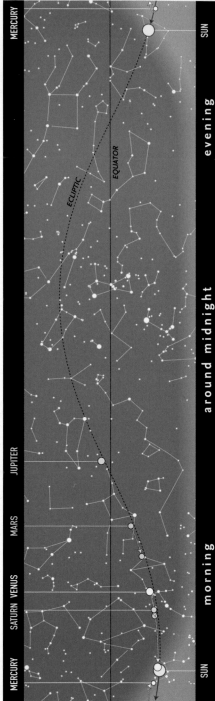

The path of the Sun and the planets along the ecliptic in January.

24

Calendar for January

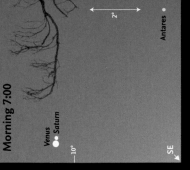

Morning 7:00

January 9 • *One day later: Venus and Saturn are very close together.*

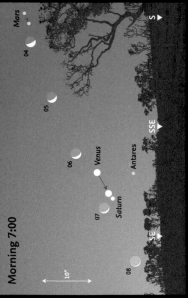

Morning 7:00

January 4-8 • *The Moon with Mars, Venus, Saturn and Antares in the morning sky.*

After midnight 1:00

Morning 7:00

January 26 • *The Moon with Regulus over the western horizon.*

January 28 • *The Moon with Jupiter in the southeastern sky, shortly after midnight.*

01–10		Quadrantid meteor shower
02	05:30	Last Quarter
02	11:53	Moon at apogee
02	22:49	Earth at perihelion (147,100,175 km = 0.9833039941 AU)
03–04		Quadrantid shower maximum
03	04:02	Spica 4.7°S of Venus
03	18:45	Mars 1.5°S of Moon
06	17:00*	Antares 6.5°S of Venus
06	23:15	Antares 9.5°S of Moon
06	23:56	Venus 3.1°S of Moon
07	04:36	Saturn 3.3°S of Moon
09	04:00*	Saturn 0.1°S of Venus
10	01:31	New Moon
14	14:05	Mercury inferior conjunction
15	02:14	Moon at perigee
16	23:26	First Quarter
20	02:40	Aldebaran 0.5°S of Moon [occultation: London 03:24; Edinburgh 03:23]
24	01:46	Full Moon
26	05:36	Regulus 2.5°N of Moon
28	01:17	Jupiter 1.4°N of Moon
30	09:10	Moon at apogee
30	12:02	Spica 5.0°S of Moon

* *These objects are close together for an extended period around this time.*

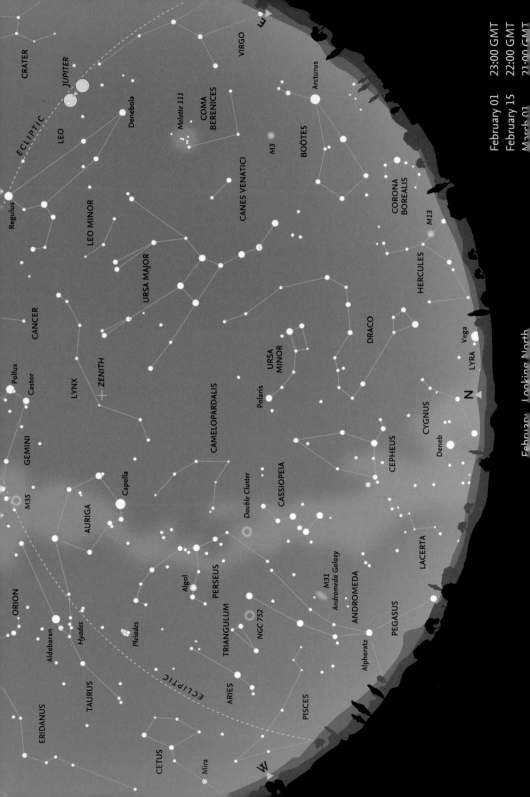

CRATER

JUPITER

ECLIPTIC

LEO

Denebola

Regulus

LEO MINOR

Melotte 111

COMA BERENICES

VIRGO

Arcturus

M3

BOÖTES

CANES VENATICI

URSA MAJOR

CORONA BOREALIS

M13

HERCULES

CANCER

LYNX

+ ZENITH

CAMELOPARDALIS

URSA MINOR

Polaris

DRACO

Vega

LYRA

Pollux
Castor

GEMINI

N

CYGNUS

Deneb

M35

Capella

AURIGA

Double Cluster

CASSIOPEIA

CEPHEUS

LACERTA

ORION

Algol

PERSEUS

M31
Andromeda Galaxy

ANDROMEDA

PEGASUS

Aldebaran

Hyades

Pleiades

TRIANGULUM

NGC 752

Alpheratz

TAURUS

ERIDANUS

ARIES

ECLIPTIC

PISCES

CETUS

Mira

W

E

February – Looking North

The months of January and February are probably the best time for seeing the section of the Milky Way that runs in the northern and western sky from **Cygnus**, low on the northern horizon, through **Cassiopeia**, **Perseus** and **Auriga** and then down through **Gemini** and **Orion**. Although not as readily visible as the denser star clouds of the summer Milky Way, on a clear night so many stars may be seen that even a distinctive constellation such as **Cassiopeia** is not immediately obvious.

The head of **Draco** is now higher in the sky and easier to recognize. Deneb, (α Cygni), the brightest star in **Cygnus**, may just be visible almost due north at midnight, early in the month, if the sky is very clear and the horizon clear of obstacles. **Vega** (α Lyrae) in **Lyra** is so low that it is difficult to see, but may become visible later in the night. The constellation of **Boötes** – sometimes described as shaped like a kite, an ice-cream cone, or the letter 'P' – with orange-tinted **Arcturus** (α Boötis), is beginning to clear the eastern horizon. Arcturus, at magnitude -0.05, is the brightest star in the northern hemisphere. The inconspicuous constellation of **Coma Berenices** is now well above the horizon in the east. The concentration of faint stars at the northwestern corner of the constellation somewhat resembles a tiny, detatched portion of the Milky Way. This is Melotte 111, an often overlooked open star cluster (which is sometimes called the Coma Cluster, but must not be confused with the important Coma Cluster of galaxies, Abell 1656, mentioned on page 41).

On the other side of the sky, in the northwest, most of the constellation of **Andromeda** is still easily seen, although **Alpheratz** (α Andromedae), the star that forms the northeastern corner of the Great Square of Pegasus – even though it is actually part of Andromeda – is becoming close to the horizon and more difficult to detect. High overhead, at the zenith, try to make out the very faint constellation of **Lynx**. It was introduced in 1687 by the famous astronomer Johannes Hevelius to fill the largely blank area between **Auriga**, **Gemini** and **Ursa Major**, and is reputed to be so named because one needed the eyes of a lynx to detect it.

A very large, and frequently ignored, open star cluster, Melotte 111, also known as the Coma Cluster, is readily visible in the eastern sky during February.

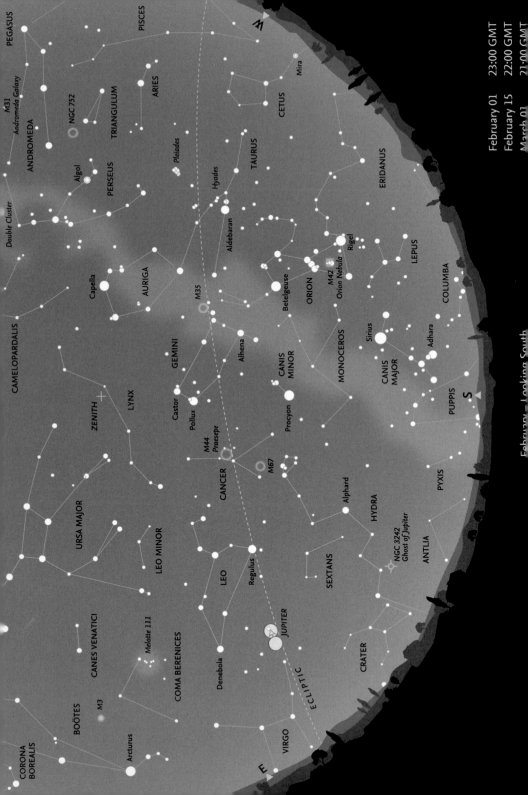

February – Looking South

February 01   23:00 GMT
February 15   22:00 GMT
March 01   21:00 GMT

February – Looking South

Apart from Orion, the most prominent constellation visible this month is **Gemini**, with its two lines of stars running southwest towards Orion. Many people have difficulty in remembering which is which of the two stars **Castor** and **Pollux**. Think of them in alphabetical order: Castor (α Geminorum), the fainter star, is closer to the North Celestial Pole. Pollux (β Geminorum) is the brighter of the two, but is farther away from the Pole. Castor is remarkable because it is actually a multiple system, consisting of no less than six individual stars.

Using Orion's belt as a guide, it points down to the southeast towards **Sirius**, the brightest star in the sky (at magnitude -1.4) in the constellation of **Canis Major**, the whole of which is now clear of the southern horizon. Forming a triangle with **Betelgeuse** in Orion and **Sirius** in Canis Major is **Procyon**, the brightest star in the small constellation of **Canis Minor**. Between Canis Major and Canis Minor is the faint constellation of **Monoceros**, which actually straddles the Milky Way, which is difficult to see in this area. Directly east of Procyon is the highly distinctive asterism of six stars that form the 'head' of **Hydra**, the largest of all 88 constellations, and which trails such a long way across

the sky that it is only in mid-March around midnight that the whole constellation becomes visible.

The constellation of Gemini. The two brightest stars are Castor and Pollux and can be found in the left part of the photograph.

The Moon's phases for February

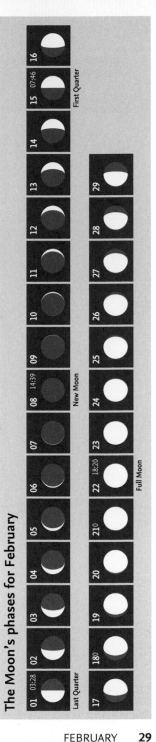

01 03:28	02	03	04	05	06	07	08 14:39	09
Last Quarter							New Moon	

10	11	12	13	14	15 07:46	16
					First Quarter	

17	18 0	19	20	21 0	22 18:20	23
					Full Moon	

24	25	26	27	28	29

February – Moon and Planets

The Moon

On February 1 the Moon, at Last Quarter, passes close to **Mars**. Two days later it is near **Saturn** and **Antares** in **Scorpius**. On February 6 it passes near **Venus** and then **Mercury** in the early-morning sky. By February 16 it is close to Aldebaran in **Taurus** (an occultation is visible at 08:04 UT from the northen Pacific). On February 22 it passes south of **Regulus** in Leo, and on February 24 south of **Jupiter**. On February 26 it is near **Spica** in **Virgo** and on February 29 near **Mars** in **Libra**.

The Planets

Mercury reaches greatest western elongation on February 7 at magnitude -0.1, close to **Venus** (magnitude -3.9) in the morning sky, but is too low to be readily seen before sunrise. **Mars** moves eastwards in **Libra** throughout February, rising in the south-east initially about 01:30 UT, and slightly earlier at the end of the month. It gradually brightens from magnitude 0.8 to 0.3. **Jupiter** is almost stationary in **Leo** at magnitude -2.4 to -2.5. **Saturn** (magnitude 0.5) is moving very slowly eastwards in **Ophiuchus**. **Uranus** (magnitude 5.9) and **Neptune** (magnitude 8.0) remain, respectively in **Pisces** and **Aquarius**.

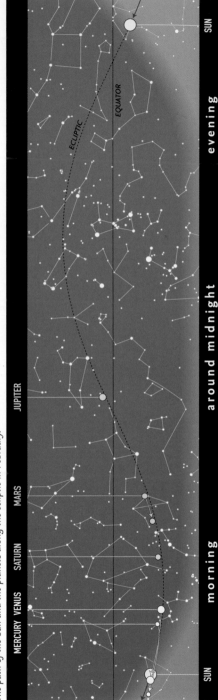

The path of the Sun and the planets along the ecliptic in February.

Calendar for February

01	03:28	Last Quarter
01	08:49	Mars 2.7°S of Moon
03	08:20	Antares 9.7°S of Moon
03	18:44	Saturn 3.5°S of Moon
06	07:31	Venus 4.3°S of Moon
06	16:47	Mercury 3.8°S of Moon
07	01:24	Mercury greatest elongation (25.6° W, mag. –0.1)
08	14:39	New Moon
10	00:19	Neptune 2.0°S of Moon
11	02:41	Moon at perigee (364,360 km)
12	13:44	Uranus 1.7°N of Moon
15	07:46	First Quarter
16	08:04	Aldebaran 0.3°S of Moon
19	17:49	Pollux 11.2°N of Moon
22	13:14	Regulus 2.5°N of Moon [Jupiter relatively nearby until 05:00 on Feb. 22]
22	18:20	Full Moon
24	04:02	Jupiter 1.7°N of Moon
26	19:33	Spica 5.1°S of Moon
27	03:28	Moon at apogee (405,383 km)
29	18:16	Mars 3.6°S of Moon

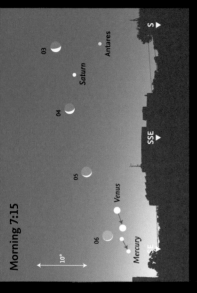

Morning 7:15

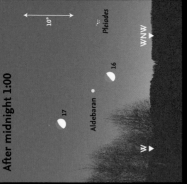

After midnight 1:00

February 3–4 • *The Moon passes Antares and Saturn. Two days later, on February 6, it is close to Venus and Mercury.*

February 16–17 • *The Moon passes Aldebaran. The conjunction is in daylight.*

Evening 18:00

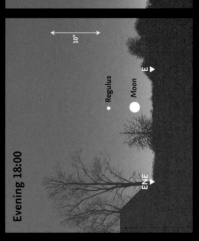

Morning 6:00

February 22 • *The Full Moon rises in the east. Regulus is almost straight above it.*

February 24 • *The Moon with Jupiter in the western sky, before sunrise. Regulus is approaching the horizon.*

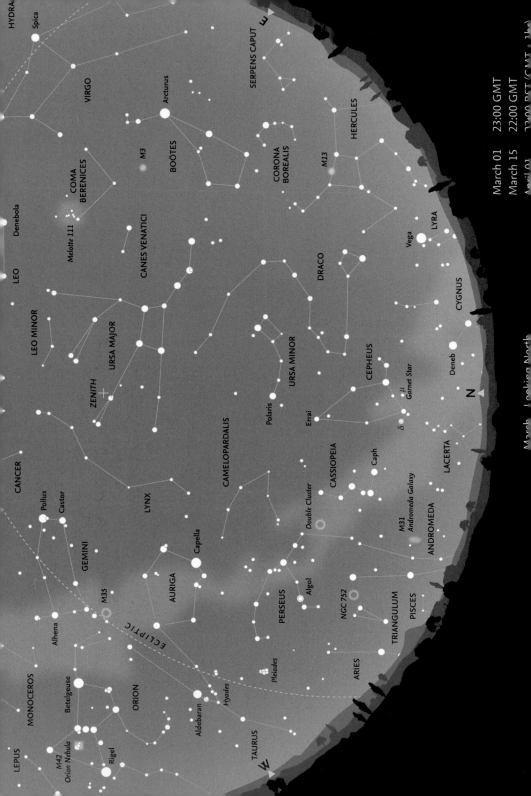

March — Looking North

March 01   23:00 GMT
March 15   22:00 GMT
April 01   22:00 BST (GMT + 1hr)

March – Looking North

In March, the Sun crosses the celestial equator on Sunday, March 20, at the vernal equinox, when day and night are of almost exactly equal length, and the season of spring is considered to have begun. (The hours of daylight and darkness change most rapidly around the equinoxes, in March and September.) It is also in March that Summer Time begins in Europe (on Sunday, March 27) so the charts show the appearance at 23:00 GMT for March 1 and 22:00 BST for April 1. (In the USA, Daylight Saving Time is introduced two weeks earlier, on Sunday, March 13.)

Early in the month, the constellation of **Cepheus** lies almost due north, with the distinctive 'W' of **Cassiopeia** to its west. Cepheus lies across the border of the Milky Way and is often described as like the gable-end of a house or a church tower and steeple. Despite the large number of stars revealed at the base of the constellation by binoculars, one star stands out because of its deep red colour. This is Mu (μ) Cephei, also known as the **Garnet Star**, because of its striking colour. It is a truly gigantic star, a red supergiant, and one of the largest stars known. It is about 2400 times the diameter of the Sun, and if placed in the Solar System would extend beyond the orbit of Saturn. (Betelgeuse, in Orion, is also a red supergiant, but it is 'only' about 500 times the diameter of the Sun.)

Another famous, and very important star in Cepheus is δ **Cephei**, which is the prototype for the class of variable stars known as Cepheids. These giant stars show a regular variation in their luminosity, and there is a direct relationship between the period of the changes in magnitude and the stars' actual luminosity. From a knowledge of the period of any Cepheid, its actual luminosity – known as its absolute magnitude – may be derived. A comparison of its apparent magnitude on the sky and its absolute magnitude enables the star's exact distance to be

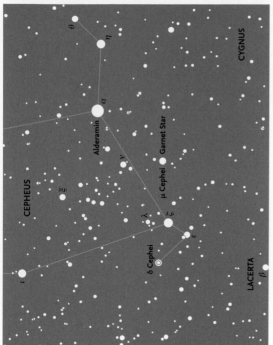

A finder chart for δ Cephei and μ Cephei (the Garnet Star). All stars brighter than magnitude 7.5 are shown.

determined. Once the distances to the first Cepheid variables had been established, examples in more distant galaxies provided information about the scale of the universe. Cepheid variables are the first 'rung' in the cosmic distance ladder.

Below Cepheus to the east (to the right), it may be possible to catch a glimpse of **Deneb** (α Cygni), just above the horizon. Slightly farther round towards the northeast, **Vega** (α Lyrae) is marginally higher in the sky. From southern Britain, Deneb is just far enough north to be circumpolar (although difficult to see in January and February because it is so low on the northern horizon). Vega, by contrast, farther south, is completely hidden during the depths of winter.

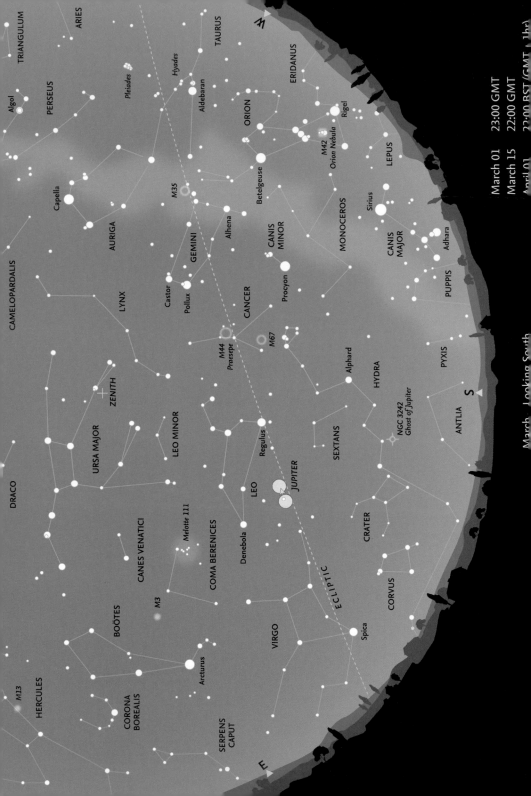

March – Looking South

March 01   23:00 GMT
March 15   22:00 GMT
April 01   22:00 BST (GMT + 1hr)

March – Looking South

Due south at 22:00 at the beginning of the month, lying between the constellations of *Gemini* in the west and *Leo* in the east, and fairly high in the sky above the head of Hydra, is the faint, and rather undistinguished zodiacal constellation of *Cancer*. Rather like the triskelion, the symbol for the Isle of Man, it has three 'legs' radiating from the centre, where there is an open cluster, M44 or *Praesepe* ('the Manger' but also known as 'the Beehive'). On a clear night this is just a hazy spot to the naked eye, but appears in binoculars as a group of dozens of individual stars.

Also prominent in March is the constellation of *Leo*, with the 'backward question mark' (or 'Sickle') of bright stars forming the head of the mythological lion. *Regulus* (α Leonis) – the 'dot' of the 'question mark' or the handle of the sickle and the brightest star in Leo – lies very close to the ecliptic and is one of the few first-magnitude stars that may be occulted by the Moon. The next occultation occurs on December 18, visible only from southern Australia and part of Antarctica. A series of occultations follows in 2017, but not until 25 July 2017 and 8 December 2017 will occultations be visible from any part of Europe.

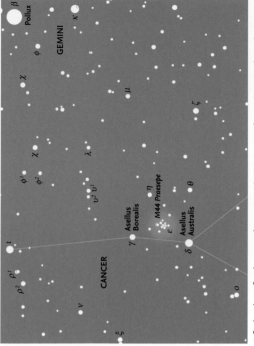

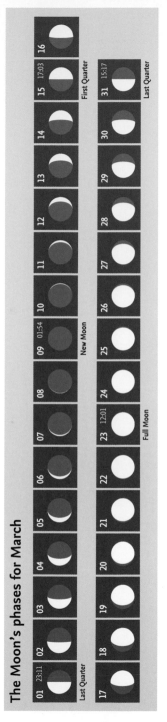

A finder chart for the open cluster M44 in Cancer. To the ancient Greeks and Romans, the two stars Asellus Borealis and Asellus Australis represented two donkeys feeding from Praesepe ('the Manger'). All stars brighter than magnitude 7.5 are shown.

The Moon's phases for March

01 23:11	02	03	04	05	06	07	08
Last Quarter							
09 01:54	10	11	12	13	14	15 17:03	16
New Moon						First Quarter	
17	18	19	20	21	22	23 12:01	24
						Full Moon	
25	26	27	28	29	30	31 15:17	
						Last Quarter	

March – Moon and Planets

The Moon

On March 1, the Last Quarter Moon passes *Antares*. It is close to both *Mars* and *Saturn* for several days shortly after midnight. On March 9, New Moon occurs at 01:54 UT and there is a total eclipse of the Sun, visible from Indonesia and neighbouring regions of the Pacific Ocean. Mid-eclipse is 01:57 UT. On March 14, a day before First Quarter, the Moon is close to *Aldebaran* in *Taurus* shortly before both set in the west. Between March 20 and 22, the Moon passes south of *Regulus* and *Jupiter*, also in the western sky. At Full Moon on March 23, there is a penumbral eclipse – when the Moon enters the Earth's outer shadow – visible only from the Pacific Ocean. Two days after Full Moon, it passes north of *Spica* in *Virgo*. Towards the end of the month, on March 28 and 29, it is again close to *Antares*, *Mars* and *Saturn*.

The Planets

Mercury is close to the Sun throughout the month, passing superior conjunction on March 23. *Venus* is too close to the Sun and too low to be visible in March. *Mars* is initially in *Libra* and at magnitude 0.3, but brightens as it crosses into *Scorpius* mid-month, reaching magnitude -0.5 by March 31. *Jupiter* is retrograding in *Leo* and comes to opposition on March 8 at magnitude -2.5. It remains in Leo and is magnitude -2.4 at the end of the month. *Saturn* is moving slowly eastwards in *Ophiuchus*, and is at magnitude 0.5–0.4 throughout the month. *Uranus* is magnitude 5.9 in *Pisces* and *Neptune* magnitude 8.0 in *Aquarius*.

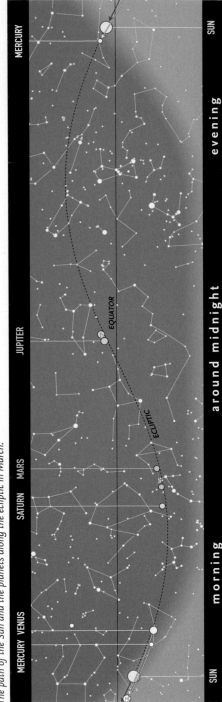

The path of the Sun and the planets along the ecliptic in March.

Calendar for March

01	16:42	Antares 9.8°S of Moon
01	23:11	Last Quarter
02	06:32	Saturn 3.6°S of Moon
07	10:53	Venus 3.5°S of Moon
08	05:08	Mercury 3.9°S of Moon
08	10:57	Jupiter at opposition (mag. –2.5)
08	12:24	Neptune 2.0°S of Moon
09	01:00	Daylight Saving Time (DST) begins in North America
09	01:54	New Moon
09	01:57	Total solar eclipse (Indonesia, Pacific Ocean)
10	07:04	Moon at perigee (359,510 km)
11	00:34	Uranus 1.9°N of Moon (North America)
14	14:07	Aldebaran 0.3°S of Moon
15	17:03	First Quarter
17	23:23	Pollux 11.3°N of Moon
20	04:30	Spring (vernal) equinox
20	19:31	Regulus 2.5°N of Moon
22	04:01	Jupiter 2.1°N of Moon
23	11:47	Penumbral lunar eclipse (Pacific)
23	12:01	Full Moon
23	20:11	Mercury at superior conjunction
25	02:17	Spica 5.1°S of Moon
25	14:17	Moon at apogee (406,125 km)
27	01:00	Summer Time begins (Europe)
28	18:46	Mars 4.2°S of Moon
28	23:43	Antares 9.8°S of Moon
29	14:36	Saturn 3.5°S of Moon
31	15:17	Last Quarter

Morning 6:30

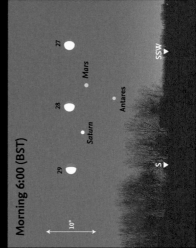

Evening 19:00

March 13–14 • *The Moon with Aldebaran high in the evening sky.*

February 29 – March 2 • *The Moon passes Mars, Antares and Saturn.*

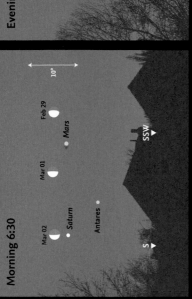

Evening 19:00

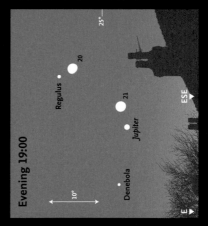

Morning 6:00 (BST)

March 20–21 • *The Moon with Regulus and Jupiter in the evening sky.*

March 27–29 • *The Moon passes Mars, Antares and Saturn in the early morning.*

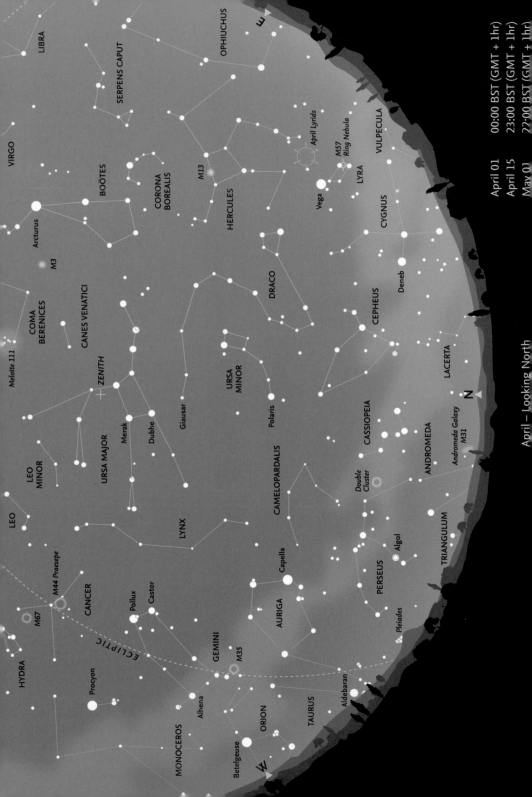

April – Looking North

LIBRA

OPHIUCHUS

SERPENS CAPUT

VIRGO

BOÖTES

Arcturus

M3

CORONA
BOREALIS

M13

HERCULES

April Lyrids

M57
Ring Nebula

Vega

LYRA

VULPECULA

COMA
BERENICES

Melotte 111

CANES VENATICI

DRACO

CYGNUS

Deneb

CEPHEUS

ZENITH

URSA MAJOR

Merak

Dubhe

Giausar

URSA
MINOR

Polaris

LACERTA

LEO

LEO
MINOR

CAMELOPARDALIS

CASSIOPEIA

ANDROMEDA

Andromeda Galaxy
M31

N

LYNX

Double
Cluster

M44 Praesepe

CANCER

M67

Pollux

Castor

Capella

Algol

PERSEUS

TRIANGULUM

HYDRA

Procyon

AURIGA

Pleiades

ECLIPTIC

GEMINI

M35

TAURUS

Aldebaran

Alhena

MONOCEROS

Betelgeuse

ORION

W

E

April 01   00:00 BST (GMT + 1hr)
April 15   23:00 BST (GMT + 1hr)
May 01   22:00 BST (GMT + 1hr)

April – Looking North

Cygnus and the brighter regions of the Milky Way are now becoming visible, rising in the northeast, with the small constellation of **Lyra** and the distinctive 'Keystone' of **Hercules** above them. This asterism is very useful for locating the bright globular cluster M13 (see map on page 53). The winding constellation of **Draco** weaves its way from the quadrilateral of stars that marks its 'head', on the border with Hercules, to end at **Giausar** (λ Draconis) between **Polaris** (α Ursae Minoris) and the 'Pointers', **Dubhe** and **Merak** (α and β Ursae Majoris, respectively). **Ursa Major** is high overhead, near the zenith. **Auriga** is still clearly seen in the northwest, but, by the end of the month, the southern portion of **Perseus** is starting to dip below the northern horizon. The very faint

constellation of **Camelopardalis** lies in the northwest between Polaris and the constellations of **Auriga** and **Perseus**.

Meteors

A minor meteor shower, the **Lyrids**, peaks on April 22–23. Although the hourly rate is not very high (about 18 meteors per hour), the meteors are fast and some leave persistent trains. This year the maximum occurs at Full Moon, so conditions are extremely unfavourable for observation. The parent object is comet C/1861 G1 (Thatcher). Another, stronger shower, the Eta Aquarids, begins to be active around April 19, and comes to maximum in May.

The constellation of Lyra is marked by the bright star Vega, which for most of Europe only dips below the northern horizon in mid-winter, with a distinctive quadrilateral of stars to its southeast.

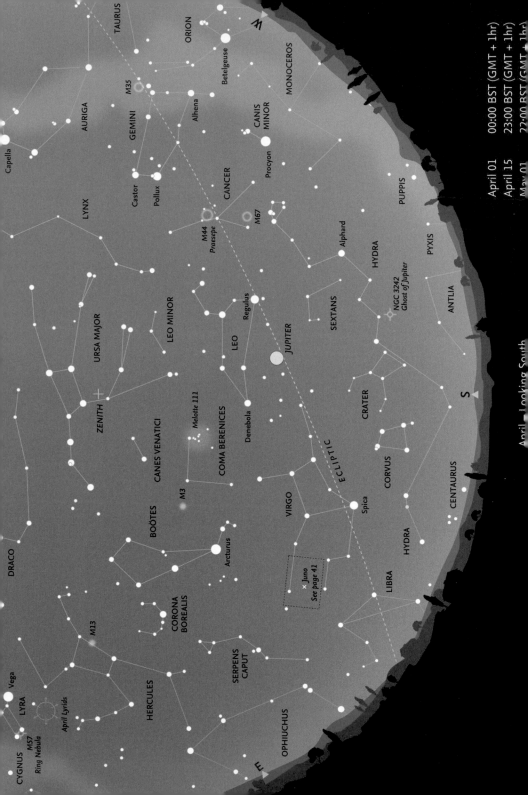

April – Looking South

April 01   00:00 BST (GMT + 1hr)
April 15   23:00 BST (GMT + 1hr)
May 01   22:00 BST (GMT + 1hr)

April – Looking South

Leo is the most prominent constellation in the southern sky in April, and vaguely looks like the creature after which it is named. **Gemini**, with **Castor** and **Pollux**, remains clearly visible in the west, and **Cancer** lies between the two constellations. To the east of Leo, the whole of **Virgo**, with **Spica** (α Virginis) its brightest star, is well clear of the horizon. Below Leo and Virgo, the complete length of **Hydra** is visible, running below Leo and Virgo, with **Alphard** (α Hydrae) half-way between Regulus and the southwestern horizon. Farther east, the two small constellations of **Crater** and the rather brighter **Corvus** lie between Hydra and Virgo.

Boötes and **Arcturus** are prominent in the eastern sky, together with the circlet of **Corona Borealis**, which is framed by Boötes and the neighbouring constellation of **Hercules**. Between Leo and Boötes lies the tiny constellation of **Coma Berenices**, most notable for being the location of the Coma Cluster of galaxies (Abell 1656). There are about 1000 galaxies in this cluster, which is located near the North Galactic Pole, where we are looking out of the plane of the Galaxy and are thus able to see deep into space. Only about ten of the brightest galaxies in the Coma Cluster are visible with the largest amateur telescopes.

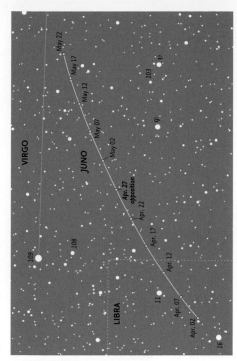

The path of minor planet (3) Juno, which is at opposition (magnitude 10.0) on April 27. Background stars are shown down to magnitude 10.5.

The Moon's phases for April

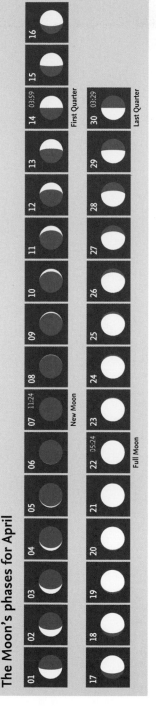

April – Moon and Planets

The Moon

New Moon is on April 7, and on the same day the Moon reaches the closest perigee of the year, at a distance of 357,163 km. Over the next three days, the Moon passes close to *Mercury* and *Aldebaran.* (There is an occultation of Aldebaran on April 10, only visible from the Atlantic Ocean.) On April 16–18 it moves past *Regulus* and *Jupiter* in *Leo.* Over the course of 6 days, from April 21 to 26 it passes (in sequence) *Spica, Mars, Antares* and *Saturn.*

The Planets

Mercury reaches greatest elongation (19.9°E) on April 18 at magnitude 0.1, after which it moves back towards the Sun and inferior conjunction on May 9. *Venus* is too low (and too close to the Sun) to be visible in April. *Mars* is almost stationary just inside *Scorpius* at the boundary with *Ophiuchus,* and brightens from magnitude -0.5 to -1.4 over the month. *Jupiter* is still slowly retrograding in Leo, fading from magnitude -2.4 to -2.3 during April. *Saturn* is in *Ophiuchus,* visible during the later part of the night, and brightens very slightly from magnitude 0.4 to 0.2. *Uranus* (magnitude 5.9) and *Neptune* (magnitude 8.0–7.9) remain in *Pisces* and *Aquarius,* respectively. On April 27, the minor planet (**3**) *Juno* reaches opposition in *Virgo* at magnitude 10.0.

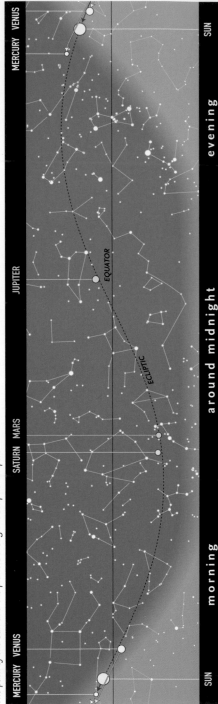

The path of the Sun and the planets along the ecliptic in April.

Calendar for April

05	00:52	Neptune 1.9°S of Moon
06	08:29	Venus 0.7°S of Moon
07	11:24	New Moon
07	13:44	Uranus 2.0°N of Moon
07	17:36	Moon at perigee (Closest of year, 357,163 km)
08	10:35	Mercury 5.2°N of Moon
10	22:27	Aldebaran 0.4°S of Moon
14	03:59	First Quarter
14	05:27	Pollux 11.2°N of Moon
16–25		Lyrid meteor shower
17	01:11	Regulus 2.6°N of Moon
18	04:46	Jupiter 2.2°N of Moon
18	13:59	Mercury greatest elongation (19.9° E, mag. 0.1)
Apr.19–May 26		Eta Aquarid meteor shower [Parent object: (1P/Halley)]
21	08:27	Spica 5.1°S of Moon
21	16:05	Moon at apogee (406,351 km)
22–23		April Lyrid shower maximum
22	05:24	Full Moon
25	04:14	Mars 4.9°S of Moon
25	05:44	Antares 9.7°S of Moon
25	19:05	Saturn 3.3°S of Moon
27	02:44	Juno at opposition (mag. 10.0)
30	03:29	Last Quarter

Evening 20:15 (BST)

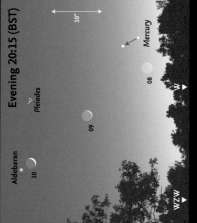

April 8–10 • *The Moon with Mercury and Aldebaran, shortly after sunset. Mercury may not be easy to see.*

Evening 21:00 (BST)

April 16–18 • *The Moon passes Regulus and Jupiter, high in the south-southeast.*

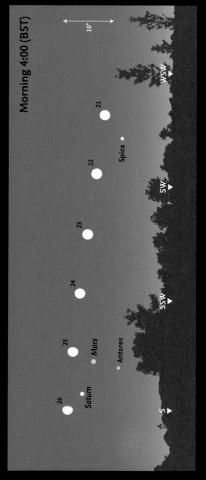

Morning 4:00 (BST)

April 21–26 • *The Moon passes Spica and, a few days later, Mars, Antares and Saturn. Mars is two magnitudes brighter than its rival Antares (the name means 'Rival of Mars').*

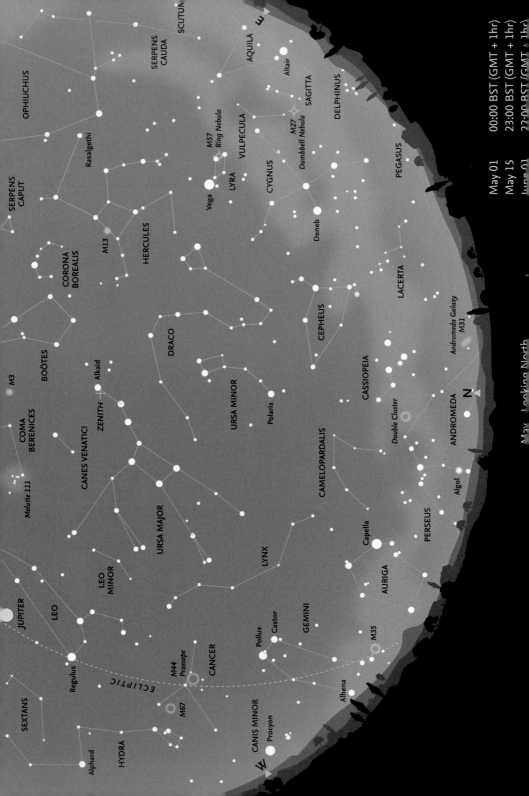

SCUTUM

SERPENS
CAUDA

OPHIUCHUS

AQUILA

Altair

SAGITTA

DELPHINUS

M57
Ring Nebula

LYRA
VULPECULA

M27
Dumbbell Nebula

Rasalgethi

SERPENS
CAPUT

Vega

CYGNUS

M13

Deneb

HERCULES

PEGASUS

CORONA
BOREALIS

LACERTA

M3

BOÖTES

DRACO

CEPHEUS

COMA
BERENICES

ZENITH   Alkaid

URSA MINOR

CASSIOPEIA

Andromeda Galaxy
M31

CANES VENATICI

Polaris

Double Cluster

N

Melotte 111

CAMELOPARDALIS

ANDROMEDA

M31

URSA MAJOR

Algol

LEO
MINOR

LYNX

PERSEUS

JUPITER

Capella

LEO

Pollux

Castor

AURIGA

Regulus

GEMINI

M44
Praesepe

CANCER

M35

M67

Alhena

SEXTANS

CANIS MINOR

ECLIPTIC

Alphard

Procyon

HYDRA

W

May, Looking North

00:00 BST (GMT + 1hr)
23:00 BST (GMT + 1hr)
22:00 BST (GMT + 1hr)

May 01
May 15
June 01

May – Looking North

Cassiopeia is now low over the northern horizon and, to its west, the southern portions of both **Perseus** and **Auriga** are becoming difficult to observe. **Gemini**, with **Castor** and **Pollux**, is sinking towards the western horizon.

In the east, two of the stars of the 'Summer Triangle', **Vega** in **Lyra** and **Deneb** in **Cygnus**, are clearly visible, and the third, **Altair** in **Aquila**, is beginning to climb above the horizon. The sprawling constellation of **Hercules** is high in the east and the brightest globular cluster in the northern hemisphere, M13, is visible to the naked eye on the western side of the 'Keystone'.

Later in the night (and in the month) the westernmost stars of **Pegasus** begin to come into view, while the stars of **Andromeda** are skimming the northeastern horizon. High overhead, **Alkaid** (η Ursae Majoris), the last star in the 'tail' of the Great Bear, is close to the zenith, while the main body of the constellation has swung round into the western sky.

Meteors

The **Eta Aquarids** are one of the two meteor showers associated with Comet 1P/Halley (the other being the **Orionids**, in October). The Eta Aquarids are not particularly favourably placed for northern-hemisphere observers, because the radiant is near the celestial equator, near the 'Water Jar' in **Aquarius**, well below the horizon until late in the night (around dawn). However, meteors may still be seen in the eastern sky even when the radiant is below the horizon. There is a radiant map for the Eta Aquarids on page 16.

Their maximum in 2016, on May 6–7, occurs when the Moon is a waning cresent, so there are very favourable observing conditions. Maximum hourly rate is about 55 per hour and a large proportion (about 25 per cent) of the meteors leave persistent trains.

Cygnus, sometimes known as the 'Northern Cross', depicts a swan flying down the Milky Way towards Sagittarius. The brightest star, Deneb (α Cygni), represents the tail, and Albireo (β Cygni) marks the position of the head, and may be found in the lower right section of the image.

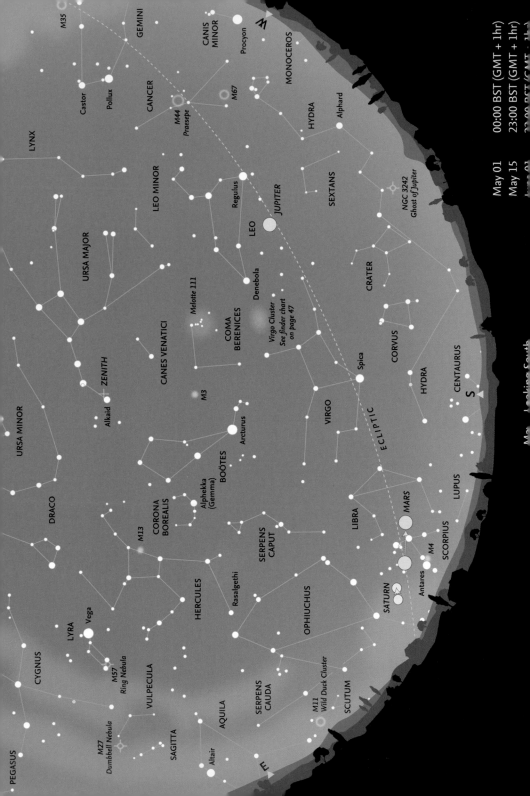

GEMINI
M35
CANIS MINOR
Procyon
MONOCEROS
Castor
Pollux
CANCER
M44
Praesepe
M67
HYDRA
Alphard
LYNX
LEO MINOR
Regulus
JUPITER
LEO
SEXTANS
URSA MAJOR
NGC 3242
Ghost of Jupiter
Denebola
Melotte 111
CRATER
COMA
BERENICES
Virgo Cluster
See finder chart
on page 47
CANES VENATICI
CORVUS
ZENITH
Alkaid
M3
Spica
CENTAURUS
HYDRA
URSA MINOR
Arcturus
VIRGO
S
BOÖTES
ECLIPTIC
DRACO
Alphekka
(Gemma)
CORONA
BOREALIS
MARS
M13
SERPENS
CAPUT
LIBRA
LUPUS
SCORPIUS
Rasalgethi
M4
HERCULES
Antares
SATURN
Vega
OPHIUCHUS
LYRA
M57
Ring Nebula
VULPECULA
CYGNUS
SERPENS
CAUDA
SCUTUM
M11
Wild Duck Cluster
AQUILA
M27
Dumbbell Nebula
SAGITTA
Altair
E
PEGASUS

May 01   00:00 BST (GMT + 1hr)
May 15   23:00 BST (GMT + 1hr)

May — Looking South

May – Looking South

Early in the night, the constellation of **Virgo**, with **Spica** (α Virginis), lies due south, with **Leo** and both **Regulus** and **Denebola** (α and β Leonis, respectively) to its west still well clear of the horizon. Later in the night, the rather faint zodiacal constellation of **Libra** becomes visible and, to its east, the ruddy star **Antares** (α Scorpii) begins to climb up over the horizon.

Virgo contains the nearest large cluster of galaxies, which is the centre of the Local Supercluster, of which the Milky Way galaxy forms part. The Virgo Cluster contains some 2000 galaxies, the brightest of which are visible in amateur telescopes.

Arcturus in **Boötes** is high in the south, with the distinctive circlet of **Corona Borealis** clearly visible to its east. The brightest star (α Coronae Borealis) is known as **Alphekka** or **Gemma**. The large constellation of **Ophiuchus** (which actually crosses the ecliptic, and is thus the 'thirteenth' zodiacal constellation) is climbing into the eastern sky. Before the constellation boundaries were formally adopted by the International Astronomical Union in 1930, the southern region of Ophiuchus was regarded as forming part of the constellation of Scorpius, which had been part of the Zodiac since antiquity.

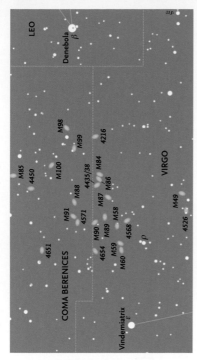

A finder chart for some of the brightest galaxies in the Virgo Cluster. All stars brighter than magnitude 8.5 are shown.

The Moon's phases for May

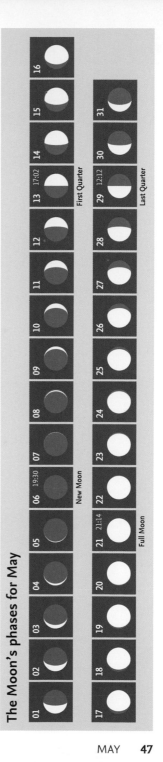

May – Moon and Planets

The Moon

Although the Moon passes close to four planets (*Neptune*, *Uranus*, *Venus* and *Mercury*) in early May, all these events, together with an occultation of *Aldebaran* on May 8, occur in daylight. On May 15, it passes below *Jupiter* in *Leo*. Three days later, on May 18, it passes north of *Spica* in *Virgo*. On May 21–23 the Moon appears close to *Mars* and *Antares* (in *Scorpius*) and *Saturn* (in *Ophiuchus*) in the southwest shortly before dawn.

The Planets

Mercury is too close to the Sun to be observed and passes inferior conjunction between the Sun and Earth on May 9. *Venus* is likewise too close to the Sun to be visible in May. *Mars* is initially visible in Scorpius, above Antares, low in the southeast and south. It retrogrades (moving westwards) rapidly into *Libra* and increases in magnitude from -1.5 to -2.0 over the course of the month, reaching opposition on May 22. *Jupiter*, in *Leo*, reaches a stationary point on May 10 and then begins direct motion (towards the east). It fades slightly from magnitude -2.3 to -2.1 over the month. *Saturn* (magnitude 0.2 to 0.0) remains in *Ophiuchus*, retrograding slowly. *Uranus* continues direct motion in *Pisces* at magnitude 5.9. Neptune is also moving slowly eastwards near λ Aquarii at magnitude 7.9. Both planets are lost in the dawn sky.

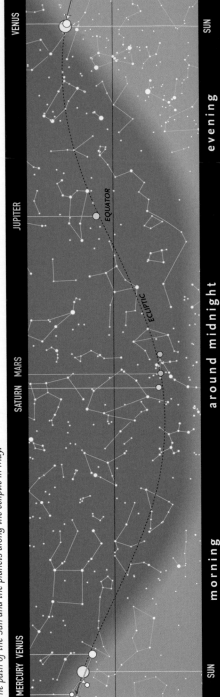

The path of the Sun and the planets along the ecliptic in May.

for May

Neptune 1.7°S of Moon
Uranus 2.2°N of Moon
Venus 2.7°N of Moon
Moon at perigee (357,827 km)
New Moon
Eta Aquarid shower maximum
Mercury 5.2°N of Moon
Aldebaran 0.5°S of Moon
Mercury at inferior conjunction
Pollux 11.0°N of Moon
First Quarter
Regulus 2.3°N of Moon
Jupiter 2.0°N of Moon
Spica 5.1°S of Moon
Moon at apogee (405,933 km)
Mars 6.0°S of Moon
Full Moon
Mars at opposition (mag. –2.1)
Antares 9.6°S of Moon
Saturn 3.2°S of Moon
Last Quarter
Neptune 1.4°S of Moon

Evening 20:50 (BST)

Betelgeuse
Pleiades
08
07
Aldebaran
10°
W
WNW

May 7–8 • The Moon with Aldebaran, above the western horizon, shortly after sunset.

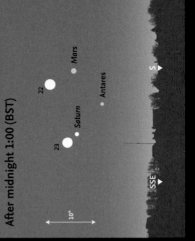

Evening 22:00 (BST)

Denebola
Regulus
13
Jupiter
14
15
10°
SW
WSW
45°

May 13 –15 • The Moon passes Regulus and Jupiter, high in the southwest.

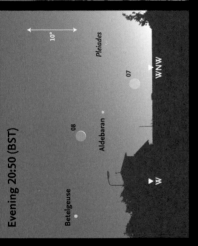

Evening 23:00 (BST)

Moon
Spica
10°
S
SSE

May 18 • The Moon with Spica in the south.

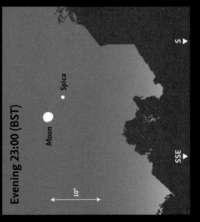

After midnight 1:00 (BST)

22
Mars
23
Saturn
Antares
10°
S
SSE

May 22–23 • The Moon passes Mars, Antares and Saturn.

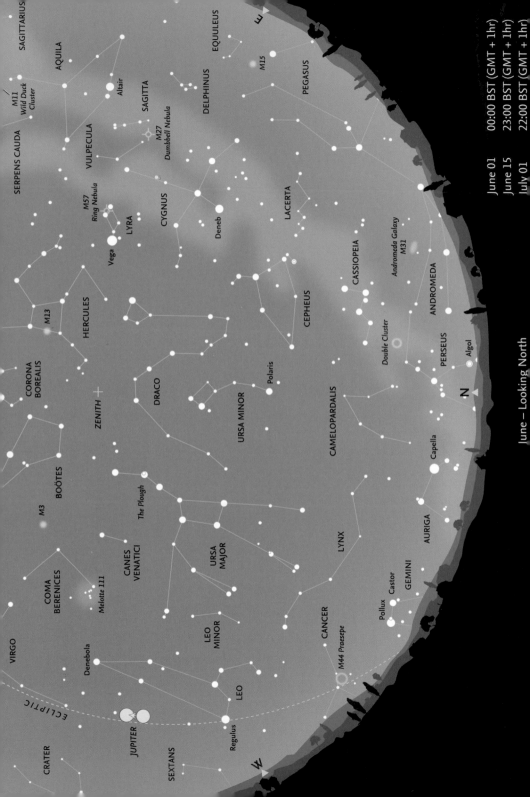

June – Looking North

SAGITTARIUS
AQUILA
EQUULEUS
M15
PEGASUS
DELPHINUS
Altair
SAGITTA
M11
Wild Duck
Cluster
SERPENS CAUDA
VULPECULA
M27
Dumbbell Nebula
LACERTA
M57
Ring Nebula
CYGNUS
LYRA
Deneb
CASSIOPEIA
Vega
Andromeda Galaxy
M31
ANDROMEDA
CEPHEUS
HERCULES
M13
Double Cluster
PERSEUS
CORONA
BOREALIS
Algol
ZENITH
DRACO
Polaris
N
URSA MINOR
BOÖTES
CAMELOPARDALIS
M3
The Plough
CANES
VENATICI
Capella
LYNX
Melotte 111
COMA
BERENICES
URSA
MAJOR
AURIGA
VIRGO
Castor
GEMINI
Pollux
LEO
MINOR
Denebola
CANCER
M44 Praesepe
ECLIPTIC
LEO
JUPITER
Regulus
CRATER
SEXTANS
W

June 01   00:00 BST (GMT + 1hr)
June 15   23:00 BST (GMT + 1hr)
July 01   22:00 BST (GMT + 1hr)

June – Looking North

As we approach the summer solstice (June 20), even in southern England and Ireland a form of twilight persists throughout the night and, farther north, in Scotland, the sky remains so light that most of the fainter stars and constellations are invisible. Even brighter stars, such as the seven stars making up the well-known asterism known as the *Plough* in *Ursa Major* may be difficult to detect except around local midnight, 00:00 UT (01:00 BST).

But there is one compensation during these light nights: even southern observers may be lucky enough to witness a display of

noctilucent clouds (NLC). These are highly distinctive clouds shining with an electric-blue tint, observed in the sky in the direction of the North Pole. They are the highest clouds in the atmosphere, occurring at altitudes of 80–85 km, far above all other clouds. They are only visible during summer nights, for about a month or six weeks on either side of the solstice, when observers are in darkness, but the clouds themselves remain illuminated by sunlight, reaching them from the Sun, itself hidden below the northern horizon.

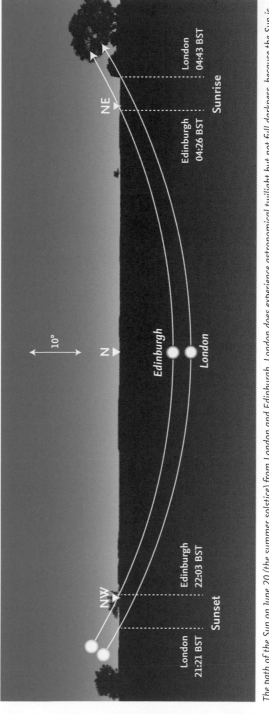

The path of the Sun on June 20 (the summer solstice) from London and Edinburgh. London does experience astronomical twilight but not full darkness, because the Sun is never more than 18° below the horizon. In Edinburgh, only nautical twilight occurs.

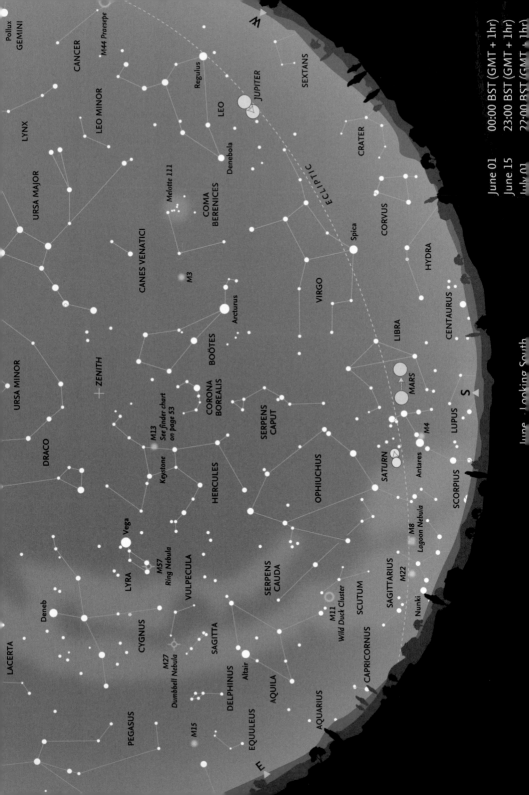

June - Looking South

June 01	00:00 BST (GMT + 1hr)
June 15	23:00 BST (GMT + 1hr)
July 01	22:00 BST (GMT + 1hr)

PEGASUS
LACERTA
EQUULEUS
M15
DELPHINUS
CYGNUS
Deneb
SAGITTA
M27 Dumbbell Nebula
VULPECULA
AQUILA
Altair
AQUARIUS
CAPRICORNUS
SCUTUM
M11 Wild Duck Cluster
SERPENS CAUDA
SAGITTARIUS
Nunki
M22
M8 Lagoon Nebula
LYRA
Vega
M57 Ring Nebula
HERCULES
M13 See finder chart on page 53
Keystone
CORONA BOREALIS
OPHIUCHUS
SERPENS CAPUT
SATURN
Antares
M4
SCORPIUS
LUPUS
S
MARS
LIBRA
BOÖTES
Arcturus
ZENITH
URSA MINOR
DRACO
CANES VENATICI
M3
COMA BERENICES
Melotte 111
VIRGO
Spica
CENTAURUS
HYDRA
CORVUS
ECLIPTIC
LEO
Denebola
Regulus
JUPITER
CRATER
SEXTANS
URSA MAJOR
LYNX
LEO MINOR
CANCER
M44 Praesepe
GEMINI
Pollux

E
N
S

June – Looking South

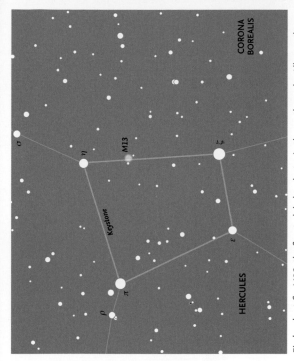

Although the persistent twilight makes observing even the southern sky difficult, the rather undistinguished constellation of **Libra** lies almost due south. The red supergiant star **Antares** – the name means the 'Rival of Mars' – in **Scorpius** is visible slightly to the east of the meridian, but the 'tail' or 'sting' remains below the horizon. Higher in the sky is the large constellation of **Ophiuchus** (the 'Serpent Bearer'), lying between the two halves of the constellation of **Serpens**: **Serpens Caput** ('Head of the Serpent') to the west and **Serpens Cauda** ('Tail of the Serpent') to the east. (Serpens is the only constellation to be divided into two distinct parts.) The ecliptic runs across Ophiuchus, and the Sun actually spends far more time in the constellation than it does in the 'classical' zodiacal constellation of Scorpius, a small area of which lies between Libra and Ophiuchus.

Higher in the southern sky, the three constellations of **Boötes**, **Corona Borealis** and **Hercules** are now better placed for observation than at any other time of the year.

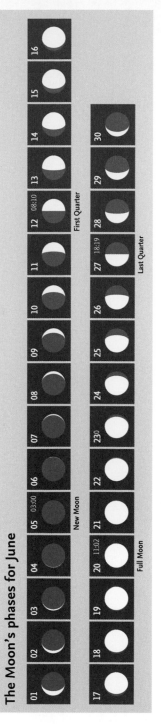

Finder chart for M13, the finest globular cluster in the northern sky. All stars down to magnitude 7.5 are shown.

The Moon's phases for June

June – Moon and Planets

The Moon

On May 4 there is an occultation of Aldebaran by the Moon, but this is not visible anywhere on Earth, occurring when it is daylight over the North Atlantic, North America and the Pacific. On May 7, the thin waxing crescent Moon passes 10.8° south of *Pollux* (β Geminorum, magnitude 1.1) in the early evening western sky. Two days later, still a crescent, it passes close to both *Regulus* and *Jupiter* in *Leo*. On June 14 it lies 5.4° north of *Spica* in *Virgo*. Over the nights of June 16–18, the waxing gibbous Moon moves past *Mars* (in *Libra*), *Antares* (in *Scorpius*) and *Saturn* (in *Ophiuchus*).

The Planets

On June 5, *Mercury* reaches greatest western elongation, but is too low in the dawn sky to be visible. The following day, June 6, *Venus* is at superior conjunction behind the Sun. *Mars* is retrograding in *Libra* and fades from magnitude -2.0 to -1.4 over the month. *Jupiter* is moving east in *Leo* in the early evening western sky, fading from magnitude -2.1 to -1.9 over the month. *Saturn* is retrograding slowly in Ophiuchus at magnitude 0.0–0.1. Uranus is still moving east in Pisces at magnitude 5.8–5.9. Neptune comes to a stationary point near λ *Aquarii* on June 13 before starting retrograde motion, and is magnitude 7.9 throughout the month.

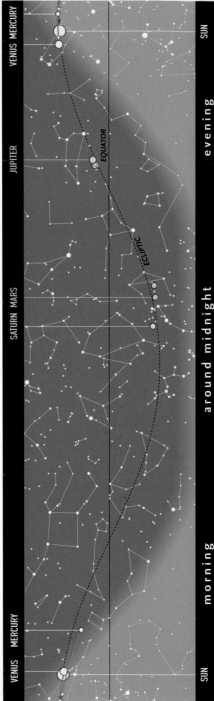

The path of the Sun and the planets along the ecliptic in June.

Calendar for June

01	14:14	Uranus 2.4°N of Moon
03	06:37	Saturn at opposition (mag. 0.0)
03	09:47	Mercury 0.7°N of Moon
03	10:55	Moon at perigee (361,140 km)
04	19:14	Aldebaran 0.5°S of Moon
05	01:07	Venus 5.0°N of Moon
05	03:00	New Moon
05	08:46	Mercury greatest elongation (24.2° W, mag. 0.5)
06	21:42	Venus at superior conjunction
07	22:40	Pollux 10.8°N of Moon
10	15:12	Regulus 2.0°N of Moon
11	19:38	Jupiter 1.5°N of Moon
12	08:10	First Quarter
14	21:14	Spica 5.4°S of Moon
15	12:01	Moon at apogee (405,024 km)
17	10:23	Mars 7.1°S of Moon
18	18:23	Antares 9.7°S of Moon
19	00:17	Saturn 3.3°S of Moon
20	11:02	Full Moon
20	22:34	Summer solstice
26	00:34	Neptune 1.2°S of Moon
27	18:19	Last Quarter
28	21:32	Uranus 2.7°N of Moon

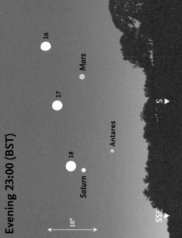

Evening 22:00 (BST)

June 7–11 • On June 7 the Moon is 10.8° south of Pollux. A few days later it passes Regulus and Jupiter.

Evening 23:00 (BST)

June 14 • The Moon with Spica in the southwest.

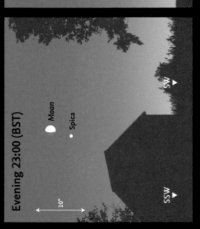

Evening 23:00 (BST)

June 16–18 • The Moon passes Mars, Antares and Saturn.

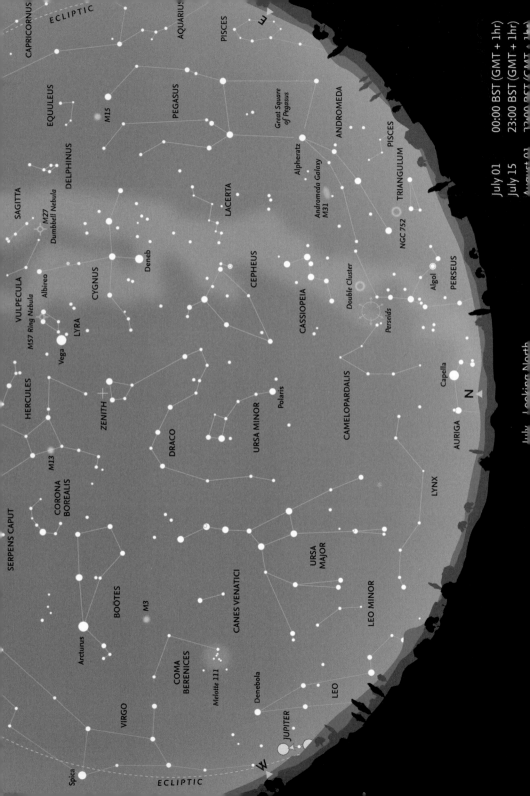

ECLIPTIC

CAPRICORNUS

AQUARIUS

PISCES

E

EQUULEUS

M15

PEGASUS

ANDROMEDA

DELPHINUS

SAGITTA

Dumbbell Nebula
M27

VULPECULA

Albireo

M57 Ring Nebula

LYRA

Vega

HERCULES

ZENITH

M13

CORONA
BOREALIS

SERPENS CAPUT

BOÖTES

M3

Arcturus

COMA
BERENICES

Melotte 111

VIRGO

Spica

ECLIPTIC

W

JUPITER

LEO

Denebola

LEO MINOR

CANES VENATICI

URSA
MAJOR

LYNX

CAMELOPARDALIS

DRACO

URSA MINOR

Polaris

CEPHEUS

CYGNUS

Deneb

LACERTA

Great Square
of Pegasus

Alpheratz

Andromeda Galaxy
M31

CASSIOPEIA

Double Cluster

Perseids

NGC 752

PISCES

TRIANGULUM

Algol

PERSEUS

Capella

N

AURIGA

July, Looking North

July 01   00:00 BST (GMT + 1hr)
July 15   23:00 BST (GMT + 1hr)
August 01   22:00 BST (GMT + 1hr)

July – Looking North

As in June, light nights and the chance of observing noctilucent clouds persist throughout July, but later in the month (and particularly after midnight) some of the major constellations begin to be more easily seen. **Capella**, the brightest star in **Auriga** (most of which is too low to be visible), is skimming the northern horizon. **Cassiopeia** is clearly visible in the northeast and **Perseus**, to its south, is beginning to climb clear of the horizon. The band of the Milky Way, from Perseus through Cassiopeia towards **Cygnus**, stretches up into the northeastern sky. If the sky is dark and clear, you may be able to make out the small, faint constellation of **Lacerta**, lying across the Milky Way between Cassiopeia and Cygnus. In the east, the stars of **Pegasus** are now well clear of the horizon, with the main line of stars forming **Andromeda** roughly parallel to the horizon in the northeast. **Alpheratz** (α Andromedae) is actually the star at the northeastern corner of the **Great Square of Pegasus. Cepheus** and **Ursa Major** are on opposite sides of **Polaris** and **Ursa Minor**, in the east and west, respectively. The head of **Draco** is very close to the zenith so the whole of this winding constellation is readily seen.

Meteors

July brings increasing meteor activity, mainly because there are several minor radiants active in the constellations of **Capricornus** and **Aquarius.** Because of their location, however, observing conditions are not particularly favourable for northern-hemisphere observers, although the first shower, the **Alpha Capricornids**, active from July 11 to August 10 (peaking July 27–28), does often produce very bright fireballs. The maximum rate, however, is only about 5 per hour. The parent body is Comet 169P/NEAT. The most prominent shower is probably that of the **Delta Aquarids**, which are active from around July 21 to August 23, with a peak on July 28–29, although even then the hourly rate is unlikely to reach 20 meteors per hour. In this case, the parent body is possibly

Comet 96P/Machholz. This year both shower maxima occur after Last Quarter, so observing conditions are quite favourable. The **Perseids**, by contrast, begin on July 13 and peak on August 12–13, when there is a waxing gibbous Moon. A chart showing the **Delta Aquarid** radiant is shown on p.16.

The New Horizons probe flew past the dwarf planet Pluto on 14 July 2015. This illustration shows an image of Pluto, the large satellite Charon and the four outer satellites, obtained earlier by the Hubble Space Telescope, together with their orbits. We hope to publish a close-up image of Pluto itself in next year's Guide.

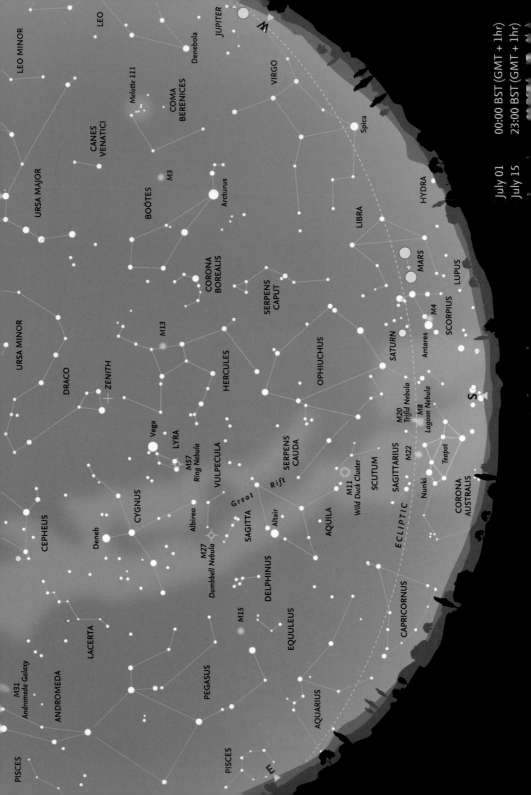

00:00 BST (GMT + 1hr)
23:00 BST (GMT + 1hr)

July 01
July 15

LEO
LEO MINOR
JUPITER
Denebola
COMA BERENICES
Melotte 111
VIRGO
CANES VENATICI
Spica
URSA MAJOR
M3
BOÖTES
Arcturus
HYDRA
LIBRA
CORONA BOREALIS
SERPENS CAPUT
MARS
URSA MINOR
M13
OPHIUCHUS
SATURN
M4
Antares
LUPUS
DRACO
ZENITH
HERCULES
SCORPIUS
S
Vega
LYRA
M20
Trifid Nebula
M8
Lagoon Nebula
M57
Ring Nebula
VULPECULA
SERPENS CAUDA
Teapot
CYGNUS
Rift
SCUTUM
M22
SAGITTARIUS
Albireo
Great
Nunki
Deneb
SAGITTA
M11
Wild Duck Cluster
ECLIPTIC
CEPHEUS
M27
Dumbbell Nebula
Altair
AQUILA
CORONA AUSTRALIS
DELPHINUS
CAPRICORNUS
PISCES
LACERTA
M15
EQUULEUS
ANDROMEDA
M31
Andromeda Galaxy
PEGASUS
AQUARIUS
PISCES
E

July – Looking South

Although part of the constellation remains hidden, this is perhaps the best time of year to see *Scorpius*, with deep red *Antares* (α Scorpii), glowing just above the southern horizon. At around midnight (UT), 01:00 BST, part of *Sagittarius*, with the distinctive asterism of the 'Teapot', and the dense star clouds of the centre of the Milky Way, are just visible in the south. With clear skies, the Great Rift – actually dust clouds that hide the more distant stars – runs down the Milky Way from Cygnus towards Sagittarius. The sprawling constellation of *Ophiuchus* lies close to the meridian for a large part of the month, separating the two halves of the constellation of *Serpens*. The western half is called *Serpens Caput* (Head of the Serpent) and the eastern part *Serpens Cauda* (Tail of the Serpent). In the east, the bright *Summer Triangle*, consisting of *Vega* in *Lyra*, *Deneb* in *Cygnus* and *Altair* in *Aquila*, begins to dominate the southern sky, as it will throughout August and into September. The small constellation of Lyra, with Vega and a distinctive quadrilateral of stars to its east and south, lies not far south of the zenith.

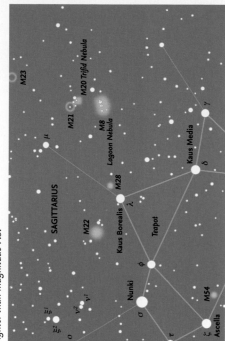

A finder chart for the gaseous nebulae M8 (the Lagoon Nebula) and M20 (the Trifid Nebula) and the globular cluster M22, all in Sagittarius. The chart shows all stars brighter than magnitude 7.5.

The Moon's phases for July

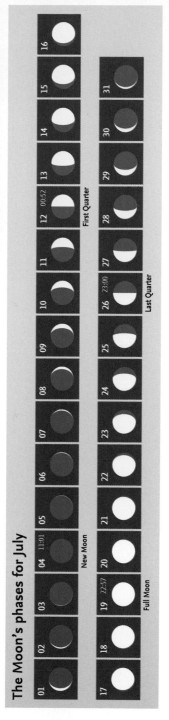

July – Moon and Planets

The Moon

On July 2, the Moon occults *Aldebaran*, although this is not visible from Britain. (A second occultation of Aldebaran, with disappearance visible from the south-western United States and parts of Central America occurs on July 29.) On July 7 and 9, the Moon passes south of *Regulus* and *Jupiter* in *Leo*. On July 14, the Moon passes north of *Mars*, and then on July 16, north of *Antares* and *Saturn*. Full Moon occurs three days later, on July 19.

The Earth

The Earth passes aphelion (the farthest point from the Sun in its orbit) on July 4 at 16:24 UT, when its distance is 1.0168 AU (150,070,274 km).

The Planets

Mercury passes superior conjunction on the far side of the Sun on July 7. Although well east of the Sun by the end of the month it is too low to be visible after sunset. *Venus* is too close to the Sun to be visible in July. *Mars* is moving east in Libra, almost reaching the border with *Scorpius* by July 31. It fades from magnitude -1.4 to -0.8 over the month. *Jupiter* is in *Leo*, close to the border with *Virgo* at magnitude -1.9 to -1.7 and, although low, may be glimpsed shortly after sunset. *Saturn* remains, slowly retrograding, in *Ophiuchus* and fades slightly from magnitude 0.1 to 0.3 over the month. *Uranus* (magnitude 5.8) has very slow direct motion as it approaches a stationary point on July 30. *Neptune*, still near λ *Aquarii* and magnitude 7.9–7.8, has started slow retrograde motion.

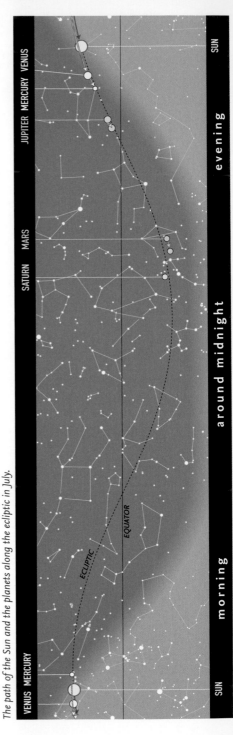

The path of the Sun and the planets along the ecliptic in July.

Calendar for July

01	06:40	Moon at perigee (365,983 km)
02	04:20	Aldebaran 0.4°S of Moon
04	11:01	New Moon
04	16:24	Earth at aphelion (150,070,274 km = 1.016750939 AU)
05	08:22	Pollux 10.7°N of Moon
07	03:24	Mercury at superior conjunction
07	23:57	Regulus 1.8°N of Moon
09	10:11	Jupiter 0.9°N of Moon
Jul.11–Aug.10		Alpha Capricornid meteor shower
12	00:52	First Quarter
12	04:40	Spica 5.6°S of Moon
13	05:24	Moon at apogee (404,269 km)
Jul.13–Aug.26		Perseid meteor shower
14	18:25	Mars 7.8°S of Moon
16	02:03	Antares 9.7°S of Moon
16	04:47	Saturn 3.4°S of Moon
19	22:57	Full Moon
Jul.21–Aug.23		Delta Aquarid meteor shower
26	04:29	Uranus 2.9°N of Moon
26	23:00	Last Quarter
27	11:37	Moon at perigee (369,662 km)
27–28		Alpha Capricornid maximum
28–29		Delta Aquarid maximum

Morning 4:30 (BST)

Hyades

Aldebaran

Moon

5°

ENE

July 2 • *The crescent Moon and Aldebaran, before sunrise. Occultation is not visible from Britain.*

Evening 22:00 (BST)

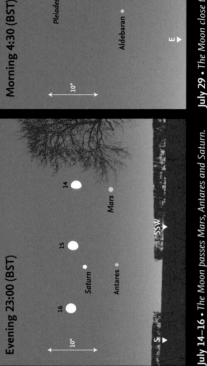

Denebola

Regulus

07

Jupiter

08

09

10°

W

WSW

July 7–9 • *The Moon passes Regulus and Denebola.*

Evening 23:00 (BST)

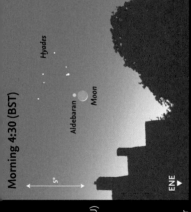

14

15

16

Mars

Saturn

Antares

10°

S

SSW

July 14–16 • *The Moon passes Mars, Antares and Saturn.*

Morning 4:30 (BST)

Pleiades

Aldebaran

10°

25°

E

ESE

July 29 • *The Moon close to Aldebaran. The*

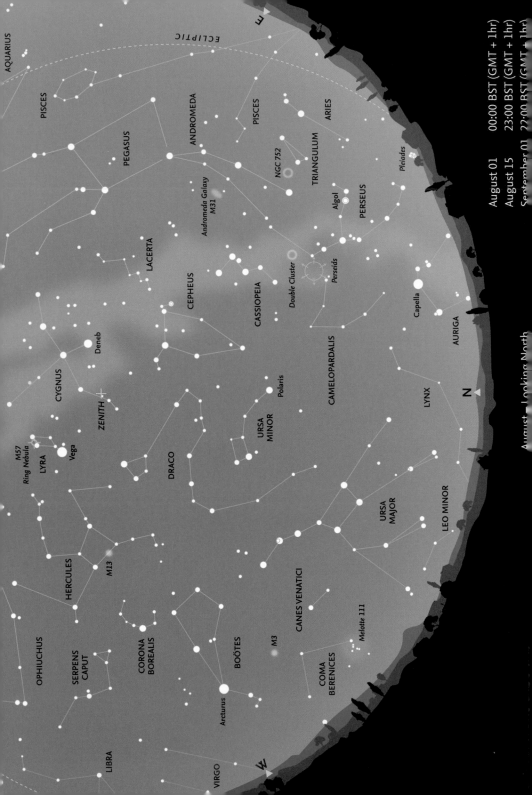

AQUARIUS

PISCES

PEGASUS

ANDROMEDA

PISCES

ARIES

LACERTA

NGC 752

TRIANGULUM

Andromeda Galaxy
M31

CEPHEUS

Algol

PERSEUS

CASSIOPEIA

Deneb

Double Cluster

Perseids

Pleiades

CYGNUS

CAMELOPARDALIS

Capella

ZENITH

AURIGA

Polaris

M57
Ring Nebula

LYRA

Vega

URSA
MINOR

LYNX

AURIGA

N

DRACO

URSA
MAJOR

HERCULES

M13

LEO MINOR

OPHIUCHUS

SERPENS
CAPUT

CORONA
BOREALIS

CANES VENATICI

Melotte 111

BOÖTES

M3

COMA
BERENICES

LIBRA

Arcturus

VIRGO

W

ECLIPTIC

E

August – Looking North

Ursa Major is now the 'right way up' in the northwest, although some of the fainter stars in the south of the constellation are difficult to see. Beyond it, *Boötes* stands almost vertically in the west, but pale orange *Arcturus* is sinking towards the horizon. Higher in the sky, both *Corona Borealis* and *Hercules* are clearly visible.

In the northeast, *Capella* is clearly visible, but most of *Auriga* still remains below the horizon. Higher in the sky, *Perseus* is gradually coming into full view and, later in the night and later in the month, the beautiful *Pleiades* cluster rises above the northeastern horizon. Between Perseus and *Polaris* lies the faint and unremarkable constellation of *Camelopardalis*.

Higher still, both *Cassiopeia* and *Cepheus* are well placed for observation, despite the fact that Cassiopeia is completely immersed in the band of the Milky Way, as is the 'base' of Cepheus. *Pegasus* and *Andromeda* are now well above the eastern horizon and, below them, the constellation of *Pisces* is climbing into view. Two of the stars in the *Summer Triangle*, *Deneb* and *Vega*, are close to the zenith high overhead.

Meteors

August is the month when one of the best meteor showers of the year occurs: the *Perseids*. This is a long shower, generally beginning about July 13 and continuing until around August 26, with a maximum in 2016 on August 12–13, when the rate may reach as high as 100 meteors per hour (and on rare occasions, even higher). In 2016, maximum is shortly after New Moon, so interference by moonlight will not be a major factor. The Perseids are debris from Comet 109P/Swift-Tuttle (the Great Comet of 1862). Perseid meteors are fast and many of the brighter ones leave persistent trains.

Comet 67P/Churyumov-Gerasimenko and Rosetta

On 13 August 2015, Comet 67P/Churyumov-Gerasimenko, accompanied by the Rosetta spaceprobe, reached perihelion. Rosetta arrived at the comet on 4 August 2014 and began its global mapping programme on September 10. An unexpected range of different terrains was discovered, and these have been colour-coded on the overall map. The various areas have been named after ancient Egyptian gods. The Philae lander was released on 12 November 2014.

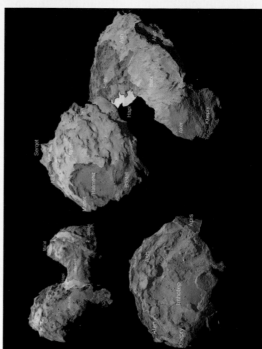

The various distinct regions of Comet 67P/Churyumov-Gerasimenko, as determined by the Rosetta spaceprobe, and shown in this global map.

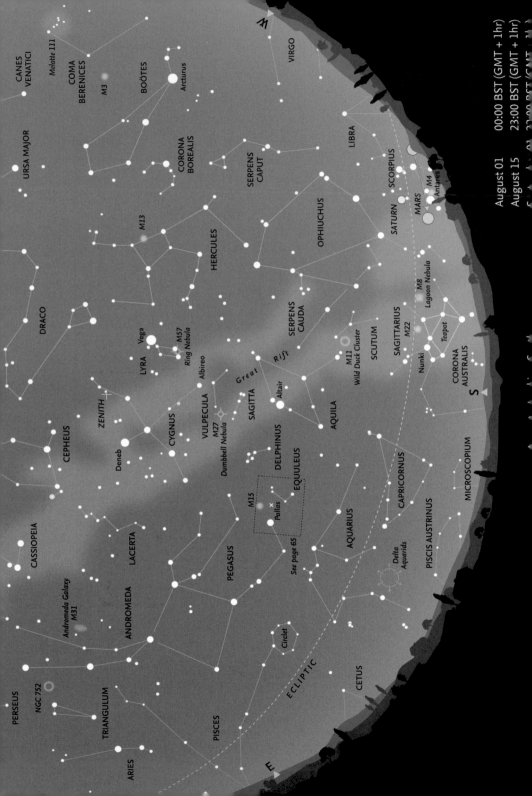

August – Looking South

The whole stretch of the summer Milky Way stretches across the sky in the south, from **Cygnus**, high in the sky near the zenith, past **Aquila**, with bright **Altair** (α Aquilae), to part of the constellation of **Sagittarius** close to the horizon, where the pattern of stars known as the 'Teapot' may be just visible. Between **Albireo** (β Cygni) and Altair lie the two small constellations of **Vulpecula** and **Sagitta**, with the latter easier to distinguish (because of its shape) from the clouds of the Milky Way. Between Sagitta and **Pegasus** to the east lie the highly distinctive five stars that form the tiny constellation of **Delphinus** (again, one of the few constellations that actually bear some resemblance to the creatures after which they are named). Below Aquila, mainly in the star clouds of the Milky Way, lies **Scutum**, most famous for the bright open cluster, M11 or the 'Wild Duck Cluster', readily visible in binoculars. To the southeast of Aquila lie the two zodiacal constellations of **Capricornus** and **Aquarius**.

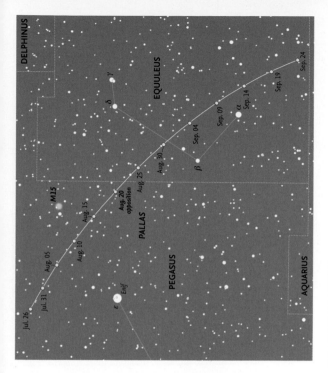

Finder chart for minor planet (2) Pallas around the time of opposition on August 25. Background stars are shown down to magnitude 9.5.

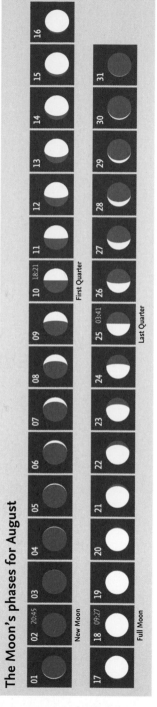

The Moon's phases for August

August – Moon and Planets

The Moon

New Moon occurs on August 2, when both it and the Sun are in *Cancer*. On August 4, it passes close to *Regulus* and *Venus*, with *Mercury* and *Jupiter* farther east, although all the planets are very close to the horizon and rather difficult to see. On August 8 in the evening sky it is close to *Spica* in *Virgo*. On August 11–12, just after First Quarter, it passes north of *Mars* and *Antares* (in *Scorpius*) and *Saturn* (in *Ophiuchus*). On August 18, the Full Moon just brushes the outer edge of the Earth's shadow, in what is technically a penumbral eclipse, although the Moon is visible only from North America and the Pacific Ocean. On August 25 (Last Quarter), the Moon passes 0.2° north of *Aldebaran*. (The star is occulted that day, but just the disappearance is visible and only from the western Pacific Ocean.)

The Planets

Mercury is initially west of the Sun, but passes superior conjunction on August 7 and then attains greatest elongation east on August 16, but is too low to be seen. *Venus* and *Jupiter* are also too close to the Sun to be visible. *Mars* is initially on the border of *Libra* and *Scorpius* but moves to the border of *Scorpius* and *Ophiuchus* at the end of the month, fading from magnitude -0.8 to -0.3. *Jupiter* remains in *Leo* at magnitude -1.7, but may perhaps be glimpsed, early in the month, low on the western horizon just before it sets. *Saturn* (magnitude 0.3–0.5) is in *Ophiuchus* throughout August. As with Mars, it remains close to the horizon throughout the month. *Uranus* (magnitude 5.8–5.7) is in the southern area of *Pisces*, and *Neptune* (magnitude 7.8) is farther west, in *Aquarius*. The minor planet *Pallas* comes to opposition on August 20 at magnitude 9.2 in *Pegasus*, close to the border with *Equuleus*.

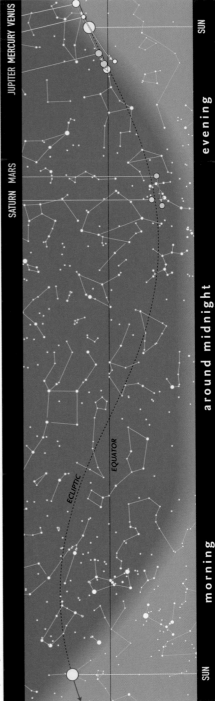

The path of the Sun and the planets along the ecliptic in August.

1	17:00	Pollux 10.7°N of Moon
2	20:45	New Moon
4	06:17	Venus 2.9°N of Moon
4	08:46	Regulus 1.7°N of Moon
4	22:11	Mercury 0.6°N of Moon
6	03:30	Jupiter 0.2°N of Moon
7	03:12	Mercury at superior conjunction
8	12:35	Spica 5.8°S of Moon
0	00:05	Moon at apogee (404,262 km)
0	18:21	First Quarter
1	21:50	Mars 8.2°S of Moon
2–13		Perseid maximum
2	10:18	Antares 9.9°S of Moon
2	11:47	Saturn 3.7°S of Moon
6	21:21	Mercury greatest elongation (27.4° E, mag. 0.2)
8	09:27	Full Moon
8	09:42	Penumbral lunar eclipse
9	12:03	Neptune 1.1°S of Moon
0	12:23	Pallas at opposition (mag. 9.2)
2	01:19	Moon at perigee (367,050 km)
2	09:57	Uranus 3.0°N of Moon
4	05:09	Mars 1.8°N of Antares
5	03:41	Last Quarter
5	16:44	Aldebaran 0.2°S of Moon
5	22:00*	Mars 4.4°S of Saturn
7	21:53	Venus 0.1°S of Jupiter
8		Alpha Aurigid first maximum
8	23:51	Pollux 10.8°N of Moon
1	16:35	Regulus 1.7°N of Moon

These objects are close together for an extended

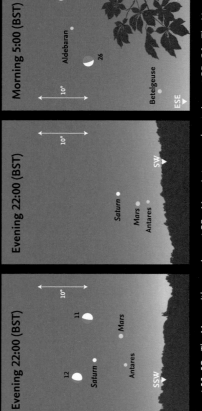

Evening 21:00 (BST)

August 4–8 • The Moon passes Regulus, Venus, Mercury, Jupiter and Spica. Regulus is very low on the horion and may not be easy to see. Even Mercury will be hard to detect.

Evening 22:00 (BST)

August 11–12 • The Moon with Mars, Antares and Saturn

Evening 22:00 (BST)

August 25 • Mars, Antares and Saturn before they set

Morning 5:00 (BST)

August 25 –26 • The Moon with Aldebaran, high in the southeast

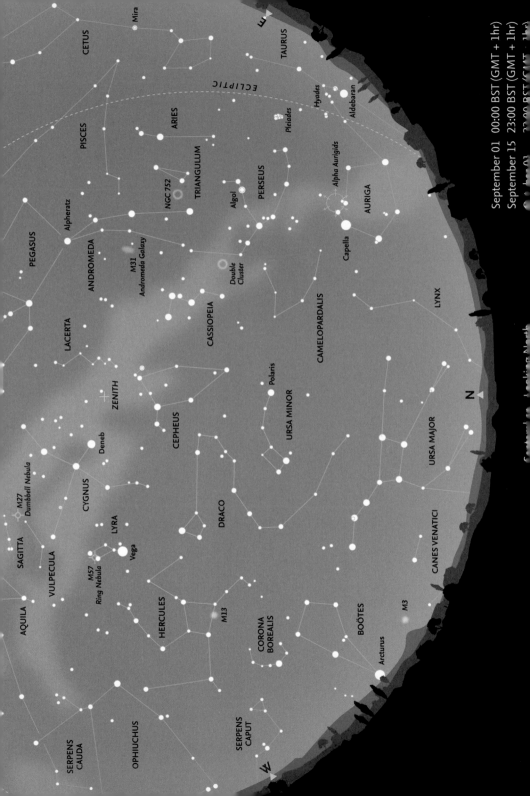

September 01 00:00 BST (GMT + 1hr)
September 15 23:00 BST (GMT + 1hr)

Centred on: Looking North

CETUS
Mira
TAURUS
Hyades
Aldebaran
ECLIPTIC
Pleiades
PISCES
ARIES
Alpha Aurigids
TRIANGULUM
PERSEUS
AURIGA
NGC 752
Algol
PEGASUS
Alpheratz
ANDROMEDA
Capella
M31
Andromeda Galaxy
Double
Cluster
LACERTA
CASSIOPEIA
CAMELOPARDALIS
LYNX
ZENITH
Polaris
CEPHEUS
URSA MINOR
CYGNUS
Deneb
N
M27
Dumbbell Nebula
DRACO
LYRA
URSA MAJOR
SAGITTA
Vega
VULPECULA
M57
Ring Nebula
HERCULES
CANES VENATICI
AQUILA
M13
CORONA
BOREALIS
BOÖTES
M3
SERPENS
CAUDA
Arcturus
OPHIUCHUS
SERPENS
CAPUT
W

September – Looking North

Ursa Major is now low in the north and to the northwest **Arcturus** and much of **Boötes** sink below the horizon later in the night and later in the month. In the northeast **Auriga** is beginning to climb higher in the sky. Later in the month, **Taurus**, with orange **Aldebaran** (α Tauri), and even **Gemini**, with **Castor** and **Pollux**, become visible in the east and northeast. Due east, **Andromeda** is now clearly visible, with the small constellations of **Triangulum** and **Aries** (the latter a zodiacal constellation) directly below it. Practically the whole of the Milky Way is visible, arching across the sky, both in the north and in the south. It is not particularly clear in Auriga, or even **Perseus**, but in **Cassiopeia** and on towards **Cygnus** the clouds of stars become easier to see. **Cepheus** is 'upside-down' near the zenith, and the head of **Draco** and **Hercules** beyond it are well placed for observation.

Meteors

After the major Perseid shower in August, there is very little shower activity in September. One minor, but very extended, shower, known as the **Alpha Aurigids**, actually has two peaks of activity. The first was on August 28 but the primary peak occurs on September 15. At either of the maxima, however, the hourly rate hardly reaches 10 meteors per hour, although the meteors are bright and relatively easy to photograph. Activity from this shower also extends into October. The **Southern Taurid** shower begins this month and, although rates are low, often produces very bright fireballs. As a slight compensation for the lack of activity, however, in September the number of sporadic meteors reaches its highest rate at any time during the year.

The twin open clusters, known as the Double Cluster, in Perseus (more formally called **h** *and* **χ Persei***) are close to the main portion of the Milky Way.*

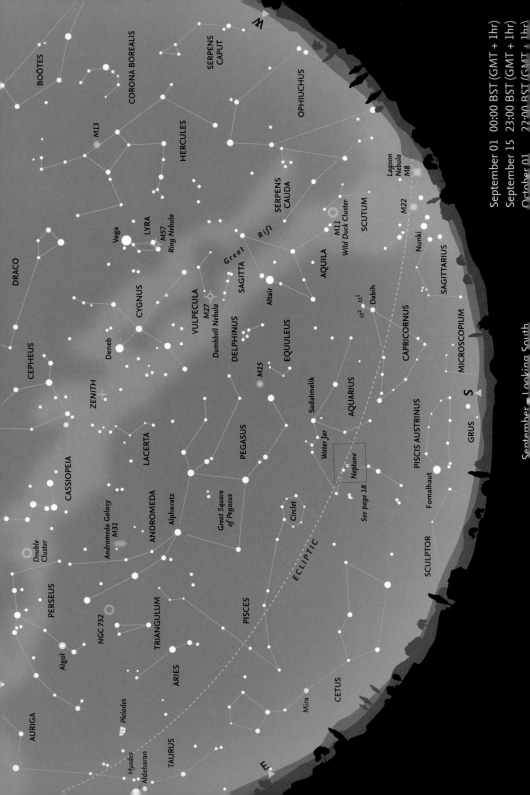

September – Looking South

September – Looking South

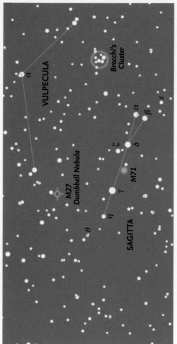

The **Summer Triangle** is now high in the southwest, with the Great Square of **Pegasus** high in the southeast. Below Pegasus are the two zodiacal constellations of **Capricornus** and **Aquarius**. In what is otherwise an unremarkable constellation, α Capricorni is actually a visual binary, with the two stars (*Prima Giedi*, α¹ Capricorni and *Secunda Giedi*, α² Capricorni) readily seen with the naked eye. *Dabih* (β Capricorni), just to the south, is also a double star, and the components are relatively easy to separate with binoculars. In Aquarius, just to the east of *Sadalmelik* (α Aquarii) there is a small asterism consisting of four stars, resembling a tiny version of Cancer, known as the 'Water Jar'. Below Aquarius is a sparsely populated area of the sky with just one bright star in the constellation of **Piscis Austrinus.** In classical illustrations, water is shown flowing from the 'Water Jar' towards bright *Fomalhaut* (α Piscis Austrini).

Another zodiacal constellation, **Pisces**, is now clearly visible to the east of Aquarius. Although faint, there is a distinctive asterism of stars, known as the 'Circlet', south of the Great Square and another line of faint stars to the east of Pegasus. Still farther down towards

A finder chart for M27, the Dumbbell Nebula, a relatively bright (magnitude 8) planetary nebula – a shell of material ejected in the late stages of a star's lifetime – in the constellation of Vulpecula. All stars brighter than magnitude 7.5 are shown.

the horizon is the constellation of **Cetus**, with the famous variable star *Mira* (ο Ceti) at its centre. When Mira is at maximum brightness (around mag. 3.5) it is clearly visible to the naked eye, but it disappears as it fades towards minimum (about mag. 9.5 or less). There is a finder chart for Mira on page 83.

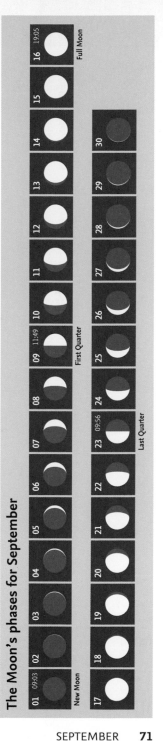

The Moon's phases for September

September – Moon and Planets

The Moon

On September 1, at New Moon there is an annular eclipse of the Sun, visible across most of Africa, Madagascar and into the Indian Ocean. From September 2–4, the Moon passes close to *Jupiter*, *Venus* and *Spica* in *Virgo*, low in the west. On September 8–9, near First Quarter, it is close to *Saturn* and *Mars* in *Ophiuchus*, and *Antares* in *Scorpius*. There is a penumbral lunar eclipse on September 16 at Full Moon. The Moon just fails to make contact with the umbra. On September 24–25, just after Last Quarter, the Moon passes very close to *Alhena* (γ Geminorum) and then *Pollux* (β Geminorum). A few days later, September 27–29, nearing New Moon, it passes near *Regulus* and very close to *Mercury* in the pre-dawn sky.

The Planets

Mercury initially invisible, reaches inferior conjunction on September 12, but is visible around greatest elongation (17.9°E) in *Leo* just before dawn on September 28. Early in the month *Venus* (magnitude -3.8 to -3.9) is in *Virgo*, but rapidly moves into *Libra* at the end of September. It is always low above the western and southwestern horizon. *Mars* is on the border of *Ophiuchus* and *Scorpius* on September 1, moving into *Sagittarius* by the end of the month, fading from magnitude -0.3 to 0.1. *Jupiter* in *Virgo* may be glimpsed in the evening sky early in September at magnitude -1.7, but is soon too close to the Sun to be seen. *Saturn* remains in *Ophiuchus* at magnitude 0.5, visible early in the night. *Uranus* (magnitude 5.7) is retrograding in *Pisces*, as is *Neptune* in *Aquarius*. Neptune comes to opposition on September 2 at magnitude 7.8. (Charts for Uranus and Neptune are given on page 18.)

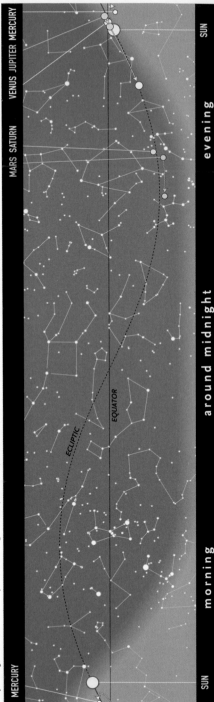

The path of the Sun and the planets along the ecliptic in September.

Calendar for September

01	09:03	New Moon
01	09:07	Annular solar eclipse (Atlantic Ocean, Africa, Madagascar, Indian Ocean)
02	21:55	Jupiter 0.4°S of Moon
02	16:38	Neptune at opposition (mag. 7.8)
03	10:31	Venus 1.1°S of Moon
04	20:23	Spica 5.8°S of Moon
06	18:45	Moon at apogee (405,055 km)
Sep.07–Nov.19		Southern Taurid meteor shower
08	18:23	Antares 10.0°S of Moon
08	21:01	Saturn 3.8°S of Moon
09	11:49	First Quarter
09	13:46	Mars 7.9°S of Moon
12	23:40	Mercury at inferior conjunction
13		Alpha Aurigid second maximum
16	19:05	Full Moon
16	18:54	Penumbral lunar eclipse (Europe, Africa, Asia)
17	23:00*	Spica 2.7°S of Venus
18	16:47	Uranus 3.0°N of Moon
18	17:00	Moon at perigee (361,896 km)
21	22:37	Aldebaran 0.2°S of Moon
22	14:21	Autumn equinox
23	09:56	Last Quarter
25	05:23	Pollux 10.7°N of Moon
27	22:56	Regulus 1.7°N of Moon
28	19:27	Mercury greatest elongation (17.9° W, mag. –0.5)
30	16:11	Jupiter 0.9°S of Moon

* These objects are close together for an extended period around this time.

Evening 20:00 (BST)

September 2–4 • The thin crescent Moon with Jupiter, Venus and Spica, shortly after sunset.

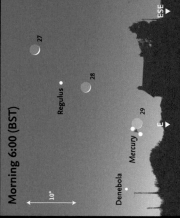

Evening 20:00 (BST)

September 8–9 • The Moon passes Antares, Saturn and Mars.

After midnight 2:00 (BST)

September 24–25 • The Moon passes Alhena and Pollux, with Castor almost 5° higher.

Morning 6:00 (BST)

September 27–29 • The Moon with Regulus and Mercury in the eastern sky, before sunrise.

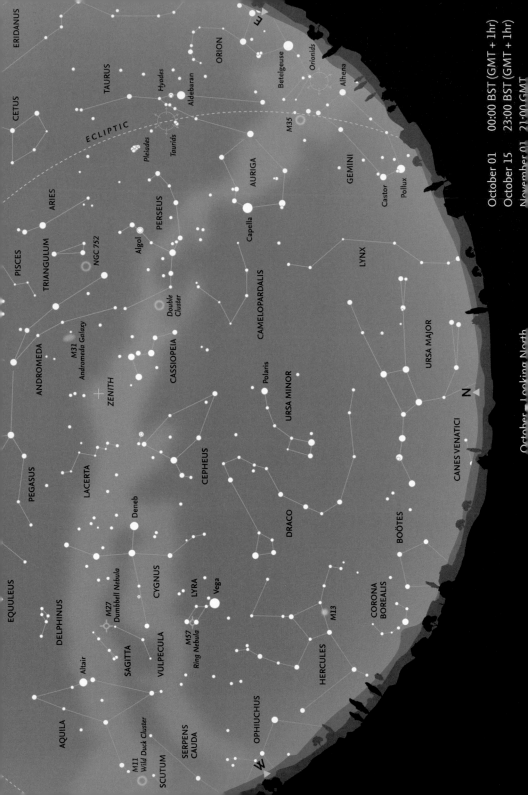

October – Looking North

October 01   00:00 BST (GMT + 1hr)
October 15   23:00 BST (GMT + 1hr)
November 01   21:00 GMT

ERIDANUS

ORION
Betelgeuse
Orionids
Alhena
Aldebaran
M35
Hyades
TAURUS
Taurids
Pleiades
ECLIPTIC
CETUS
ARIES
PISCES
TRIANGULUM
NGC 752
Algol
PERSEUS
Double
Cluster
CASSIOPEIA
ANDROMEDA
M31
Andromeda Galaxy
ZENITH
Capella
AURIGA
CAMELOPARDALIS
GEMINI
Castor
Pollux
LYNX
URSA MAJOR
Polaris
URSA MINOR
N
CANES VENATICI
CEPHEUS
PEGASUS
LACERTA
DRACO
BOÖTES
Deneb
CYGNUS
EQUULEUS
DELPHINUS
M27
Dumbbell Nebula
SAGITTA
VULPECULA
LYRA
Vega
M57
Ring Nebula
HERCULES
M13
CORONA
BOREALIS
Altair
AQUILA
SERPENS
CAUDA
OPHIUCHUS
M11
Wild Duck Cluster
SCUTUM
W
E

October – Looking North

Ursa Major is grazing the horizon in the north, while high overhead are the constellations of **Cepheus**, **Cassiopeia** and **Perseus**, with the Milky Way between Cepheus and Cassiopeia at the zenith. **Auriga** is now clearly visible in the east, as is **Taurus** with the **Pleiades**, **Hyades** and orange **Aldebaran**. Also in the east, **Orion** and **Gemini** are starting to rise clear of the horizon.

The constellations of **Boötes** and **Corona Borealis** are now essentially lost to view in the northwest, and **Hercules** is also descending towards the western horizon. The three stars of the Summer Triangle are still clearly visible, although **Aquila** and **Altair** are beginning to approach the horizon in the west. Towards the end of the month (October 30) Summer Time ends in Britain with Britain reverting to Greenwich Mean Time and Europe to Central European Time.

Meteors

The **Orionids** are the major, fairly reliable meteor shower active in October. Like the May **Eta Aquariid** shower, the Orionids are associated with Comet 1P/Halley. During this second pass through the stream of particles from the comet, slightly fewer meteors are seen than in May, but conditions are more favourable for northern observers. In both showers the meteors are very fast, and many leave persistent trains. Although the Orionid maximum is quoted as October 21–22, in fact there is a very broad maximum, lasting about a week from October 20–27, with hourly rates around 25. Occasionally rates are higher (50–70 per hour).

The faint shower of the **Southern Taurids** (often with bright fireballs) peaks on October 23–24. Towards the end of the month, another shower (the **Northern Taurids**) begins to show activity, which peaks early in November. The parent comet for both Taurid showers is Comet 2P/Encke.

A geological map of the dwarf planet (1) Ceres from observations made by the Dawn spaceprobe.

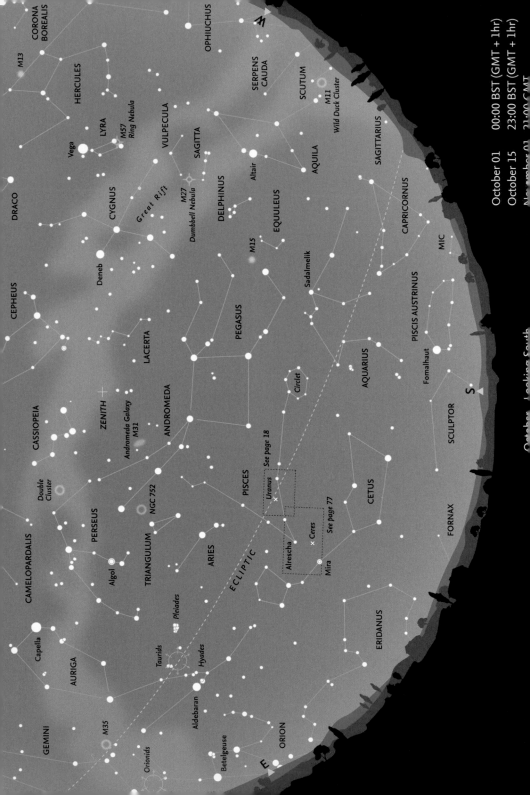

CORONA
BOREALIS

M13

HERCULES

DRACO

LYRA

Vega

M57
Ring Nebula

CYGNUS

Great Rift

Deneb

M27
Dumbbell Nebula

VULPECULA

SAGITTA

DELPHINUS

EQUULEUS

Altair

AQUILA

M15

OPHIUCHUS

SERPENS
CAUDA

SCUTUM

M11
Wild Duck Cluster

SAGITTARIUS

CAPRICORNUS

Sadalmelik

MIC

AQUARIUS

PISCIS AUSTRINUS

Fomalhaut

S

SCULPTOR

FORNAX

CETUS

Circlet

PISCES

See page 18

Uranus
×

Alrescha

× *Ceres*

See page 77

× Mira

ERIDANUS

CEPHEUS

ZENITH

Andromeda Galaxy
M31

ANDROMEDA

CASSIOPEIA

LACERTA

PEGASUS

*Double
Cluster*

NGC 752

PERSEUS

TRIANGULUM

ARIES

ECLIPTIC

Algol

CAMELOPARDALIS

Pleiades

Taurids

Aldebaran

Hyades

ORION

E

Betelgeuse

Capella

AURIGA

GEMINI

M35

Orionids

W

October – Looking South

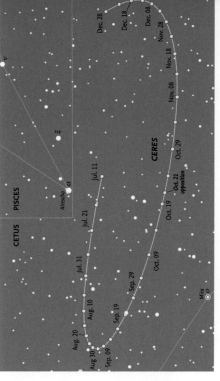

The Great Square of *Pegasus* dominates the southern sky, framed by the two chains of stars that form the constellation of *Pisces*, together with *Alrescha* (α Piscium) at the point where the two lines of stars join. Also clearly visible is the constellation of *Cetus*, below Pegasus and Pisces. Although *Capricornus* is beginning to disappear, *Aquarius* to its east is well placed in the south, with solitary *Fomalhaut* and the constellation of *Piscis Austrinus* beneath it, close to the horizon.

The main band of the Milky Way and the Great Rift runs down from *Cygnus*, through *Vulpecula*, *Sagitta* and *Aquila* towards the western horizon. *Delphinus* and the tiny, unremarkable constellation of *Equuleus* lie between the band of the Milky Way and Pegasus. *Andromeda* is clearly visible high in the sky to the southeast, with the small constellation of *Triangulum* and the zodiacal constellation of *Aries* below it. *Perseus* is high in the east, and by now the *Pleiades* and *Taurus* are well clear of the horizon. Later in the night, and later in the month, *Orion* rises in the east, a sign that the autumn season has arrived and of the steady approach of winter.

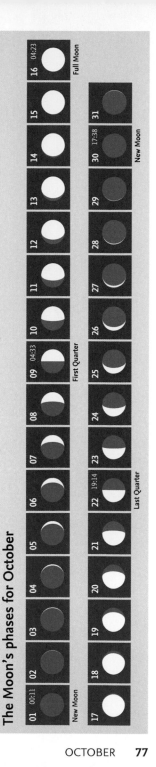

Finder chart for dwarf planet (1) Ceres for July to December 2016. Background stars are shown down to magnitude 9.0.

The Moon's phases for October

October – Moon and Planets

The Moon

New Moon occurs twice this month, on October 1 and 30. On October 2, the extremely thin crescent passes 5.8°N of **Spica** in **Virgo**. The following day it is 5.0° north of **Venus** in **Libra** and may be glimpsed just before Venus sets. On October 19, the waning gibbous Moon passes 0.3° north of **Aldebaran**. (There is an occultation of the star visible from the central and southern areas of the United States.) On October 25 the Moon is 1.6° south of **Regulus** in **Leo**. On October 28, it passes 1.4° north of **Jupiter** near **Porrima** (γ Virginis). On October 31 the Moon reaches the greatest apogee of the year (a distance of 406,662 km from the Earth).

The Planets

On October 11 **Mercury** may be glimpsed before dawn 0.8° north of **Jupiter** near **Porrima** (γ Virginis). **Venus** is very low but may be visible after sunset at the end of the month, close to **Antares** and **Saturn. Mars,** at magnitude 0.1–0.4, moves eastwards through **Sagittarius** over the month. **Jupiter** (magnitude -1.7) is close to the Sun but may be seen just before dawn early in the month. **Saturn** (magnitude 0.5) remains in **Ophiuchus** in the southwestern sky shortly after sunset. **Uranus** comes to opposition in **Pisces** at magnitude 5.7 on October 10 and continues retrograding as shown in the chart on page 18. **Neptune,** following opposition on September 2, also retrogrades in **Aquarius** at magnitude 7.8–7.9. On October 21, dwarf planet **(1) Ceres** comes to opposition at magnitude 7.4. A finder chart is given on page 77.

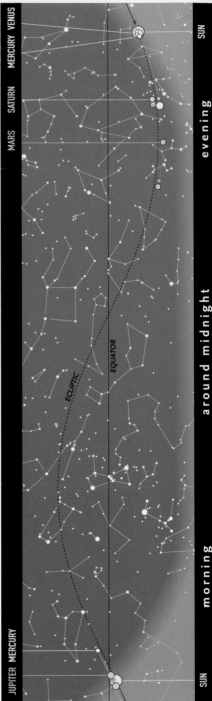

The path of the Sun and the planets along the ecliptic in October.

Calendar for October

01	00:11	New Moon
02	03:29	Spica 5.8°S of Moon
03	17:29	Venus 5.0°S of Moon
Oct.04–Nov.14		Orionid meteor shower
04	11:03	Moon at apogee (406,096 km)
06	01:37	Antares 9.9°S of Moon
06	07:43	Saturn 3.8°S of Moon
08	12:08	Mars 7.0°S of Moon
09	04:33	First Quarter
11	11:04	Mercury 0.8°N of Jupiter
15	10:43	Uranus at opposition (mag. 5.7)
16	04:23	Full Moon
16	23:34	Moon at perigee (357,861 km)
Oct.19–Dec.10		Northern Taurid meteor shower
19	06:40	Aldebaran 0.3°S of Moon
21	05:16	Ceres at opposition (mag. 7.4)
21–22		Orionid shower maximum
22	11:14	Pollux 10.6°N of Moon
22	19:14	Last Quarter
23–24		Southern Taurid shower maximum
25	04:26	Regulus 1.6°N of Moon
26	04:00*	Antares 3.1°S of Venus
28	09:35	Jupiter 1.4°S of Moon
29	09:44	Spica 5.7°S of Moon
30	01:00	Summer Time ends (Europe)
30	08:00*	Saturn 3.0°N of Venus
30	17:38	New Moon
31	19:29	Moon at apogee
		(Farthest of year, 406,662 km)

These objects are close together for an extended

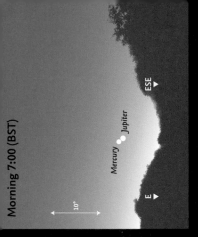

Evening 19:00 (BST)

October 3 • *The crescent Moon with Venus, shortly after sunset.*

Morning 7:00 (BST)

October 11 • *Mercury (magnitude –0.1) and Jupiter (magnitude –1.7) in the early morning sky.*

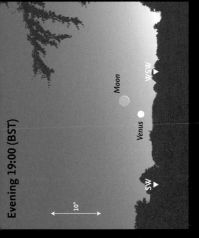

Morning 5:00 (BST)

October 25 • *The Moon with Regulus in the early*

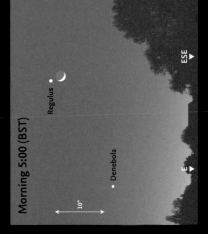

Evening 18:00 (BST)

October 26 • *Venus with Saturn and Antares over the*

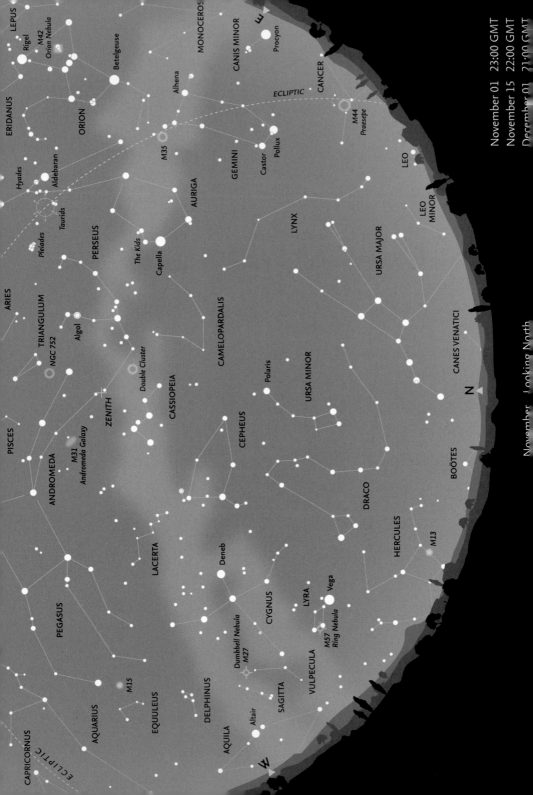

November 01   23:00 GMT
November 15   22:00 GMT
December 01   21:00 GMT

November Looking North

November – Looking North

Aquila has now disappeared below the horizon, but two of the stars of the Summer Triangle, **Vega** in **Lyra** and **Deneb** in **Cygnus**, are still visible in the west. The head of **Draco** is now low in the northwest and only a small portion of **Hercules** remains above the horizon. The Milky Way arches overhead, with the denser star clouds in the west and the less heavily populated region through **Auriga** and **Monoceros** in the east. High overhead, **Cassiopeia** is near the zenith and **Cepheus** has swung round to the northwest, while Auriga is now high in the northeast. **Gemini**, with **Castor** and **Pollux**, is well clear of the eastern horizon, and even **Procyon** (α Canis Minoris) is just climbing into view almost due east.

The constellation of Auriga, with Capella its brightest star (top right), and the distinctive triangle of 'The Kids' to its west (centre right) are prominent features in the northern sky. Capella itself is circumpolar for latitude 50° north and skims the horizon during the summer, but never actually disappears below it.

Meteors

The **Northern Taurid** shower, which began in mid-October, reaches maximum – although with only a low rate of about five meteors per hour – on November 11–12, but gradually trails off, ending around December 10. There is an apparent 7-year periodicity in fireball activity, and 2016 is unlikely to be another peak year. Far more striking, however, are the **Leonids,** which have a short period of activity (November 5–30), with maximum on November 17–18. This shower is associated with Comet 55P/Tempel-Tuttle and has shown extraordinary activity on various occasions with many thousands of meteors per hour. High rates were seen in 1999, 2001 and 2002 (reaching about 3000 meteors per hour) but have fallen dramatically since then. The rate in 2016 is likely to be about 15 per hour. These meteors are the fastest shower meteors recorded (about 70 km per second) and often leave persistent trains. The shower is very rich in faint meteors. In 2016, maximum is a few days after Full Moon, with a waning gibbous Moon in the north of **Orion**, so conditions are not particularly favourable. Observing conditions are normally best after midnight, but in this case early evening watches may detect more meteors.

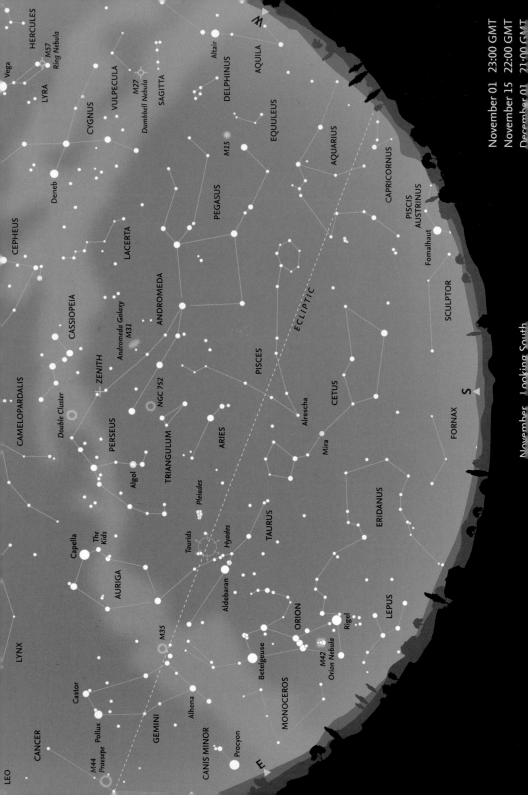

November 01 23:00 GMT
November 15 22:00 GMT
December 01 21:00 GMT

November Looking South

November – Looking South

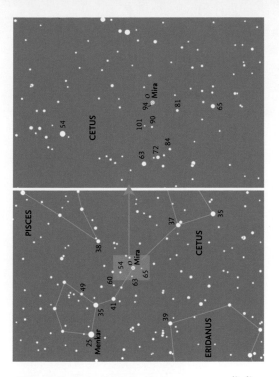

Orion has now risen above the eastern horizon, and part of the long, straggling constellation of *Eridanus* (which begins near *Rigel*) is visible to the west of Orion. Higher in the sky, *Taurus*, with the *Pleiades* cluster, and orange *Aldebaran* are now easy to observe. To their west, both *Pisces* and *Cetus* are close to the meridian. In the southwest, *Capricornus* has slipped below the horizon, but *Aquarius* remains visible. Even farther west, *Altair* may be seen early in the night, but most of *Aquila* has already disappeared from view. *Delphinus*, together with *Sagitta* and *Vulpecula* in the Milky Way, will soon vanish for another year. Both *Pegasus* and *Andromeda* are easy to see, and one of the lines of stars that make up Andromeda finishes close to the zenith, which is also close to one of the outlying stars of *Perseus*, high in the east.

Finder and comparison charts for Mira (o Ceti). The chart on the left shows all stars brighter than magnitude 6.5. The chart on the right shows stars down to magnitude 10.0. The comparison star magnitudes are shown without the decimal point.

The Moon's phases for November

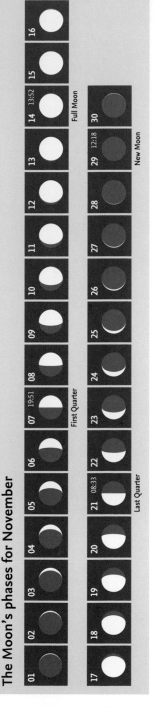

November – Moon and Planets

The Moon

In the early evening of November 2 the Moon is close to both *Venus* and *Saturn* in the southern portion of *Ophiuchus*. Over the next few days it crosses *Sagittarius* and on November 6 is close to *Mars* just inside *Capricornus*. In North America, Daylight Saving Time ends on November 6. On November 15 the Moon passes 0.5° north of *Aldebaran* and there is yet another occultation, visible only from *Asia*. On November 21 it is 1.3° south of *Regulus* in *Leo*. On November 25–26, shortly before New Moon, it is in the vicinity of *Jupiter* and *Spica* in *Virgo*.

The Planets

Mercury is too close to the Sun and too low to be visible during the month. *Venus* is visible in the early evening, moving rapidly from *Ophiuchus* to western *Sagittarius* during November, brightening from magnitude -4.0 to -4.2. *Mars* is visible in the south as it moves eastwards from *Sagittarius* into *Capricornus*, fading slightly (magnitude 0.4–0.6) over the month. *Jupiter* is visible in the early morning sky, at magnitude -1.7 to -1.8, moving slowly eastwards in *Virgo*. *Saturn* is still in *Ophiuchus*, but is too close to the Sun to be observed. *Uranus* is still retrograding in Pisces at magnitude 5.7 and *Neptune* in Aquarius at magnitude 7.9 is moving very slowly as it nears a stationary point (in December).

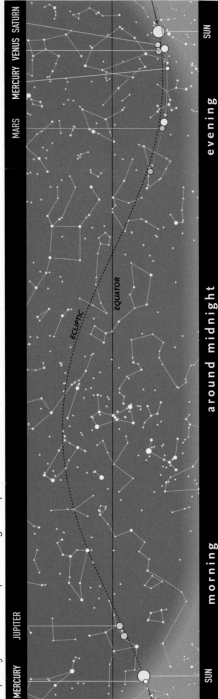

The path of the Sun and the planets along the ecliptic in November.

Calendar for November

02	07:55	Antares 9.7°S of Moon
02	19:18	Saturn 3.7°S of Moon
03	04:16	Venus 6.8°S of Moon
05–30		Leonid meteor shower
06	01:00	Daylight Saving Time ends (North America)
06	12:08	Mars 5.3°S of Moon
07	19:51	First Quarter
11–12		Northern Taurid shower maximum
14	11:21	Moon at perigee (356,5098 km)
14	13:52	Full Moon
15	17:12	Aldebaran 0.5°S of Moon
17–18		Leonid shower maximum
18	19:07	Pollux 10.4°N of Moon
21	08:33	Last Quarter
21	10:33	Regulus 1.3°N of Moon
25	01:49	Jupiter 1.9°S of Moon
25	15:36	Spica 5.9°S of Moon
27	20:08	Moon at apogee (406,554 km)
29	12:18	New Moon
29	13:56	Antares 9.7°S of Moon

Evening 17:00

November 2–6 • The Moon passes Saturn (magnitude 0.5), Venus (magnitude -4.1), Nunki (σ Sagittarii, magnitude 2.1) and Mars (magnitude 0.5).

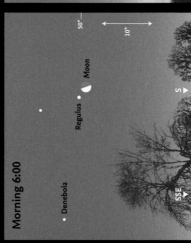

Morning 6:00

November 21 • The Last Quarter Moon with Regulus high in the southern sky, in the early morning.

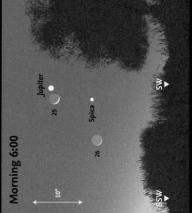

Morning 6:00

November 25–26 • The Moon with Jupiter and Spica, early in the morning.

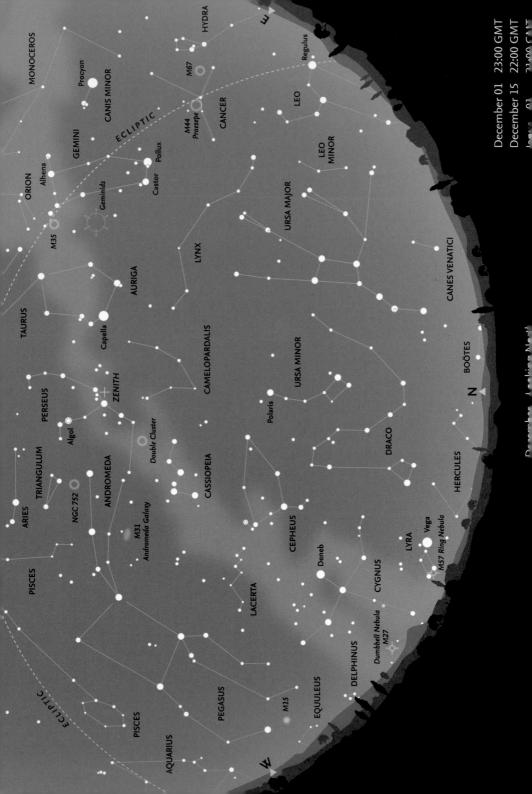

MONOCEROS

HYDRA

E

Procyon

CANIS MINOR

Regulus

M67

GEMINI

ECLIPTIC

M44
Praesepe

CANCER

LEO

Pollux

Castor

Geminids

LEO
MINOR

ORION

Alhena

URSA MAJOR

M35

LYNX

TAURUS

AURIGA

CANES VENATICI

Capella

CAMELOPARDALIS

URSA MINOR

BOÖTES

Polaris

N

ZENITH

PERSEUS

Algol

Double Cluster

DRACO

TRIANGULUM

ANDROMEDA

CASSIOPEIA

HERCULES

ARIES

NGC 752

CEPHEUS

Vega

LYRA

M31
Andromeda Galaxy

M57 Ring Nebula

PISCES

Deneb

CYGNUS

LACERTA

PEGASUS

Dumbbell Nebula
M27

PISCES

M15

DELPHINUS

EQUULEUS

AQUARIUS

W

ECLIPTIC

December Looking North

December – Looking North

Ursa Major has now swung round and is starting to 'climb' in the east. The fainter stars in the southern part of the constellation are now fully in view. The other bear, **Ursa Minor**, 'hangs' below **Polaris** in the north. Directly above it is the faint constellation of **Camelopardalis**, with the other inconspicuous circumpolar constellation, **Lynx**, to its east. **Vega** (α Lyrae) is skimming the horizon in the northwest, but **Deneb** (α Cygni) and most of **Cygnus** remain visible farther west. In the east, **Regulus** (α Leonis) and the constellation of **Leo** are beginning to rise above the horizon. **Cancer** stands high in the east, with **Gemini** even higher in the sky. **Perseus** is at the zenith, with **Auriga** and **Capella** between it and Gemini. Because it is so high in the sky, now is a good time to examine the star clouds of the fainter portion of the Milky Way, between **Cassiopeia** in the west to Gemini and **Orion** in the east.

Meteors

There is one significant meteor shower in December (the last major shower of the year). This is the **Geminid** shower, which is visible over the period December 4–16 and comes to maximum on December 13–14 with a young cresent Moon. It is one of the most active showers of the year, and in some years is the most active, with a peak rate of around 100 meteors per hour. It is the one major shower that shows good activity before midnight. The meteors have been found to have a much higher density than other meteors (which are derived from cometary material). It was eventually established that the Geminids and the asteroid Phaeton had similar orbits. So the Geminids are assumed to consist of denser, rocky material. They are slower than most other meteors and often appear to last longer. The brightest often break up into numerous luminous fragments that follow similar paths across the sky. There is a second shower: the **Ursids**, active December 17–23, peaking on December 21–22, with rate at maximum

of 5–10, occasionally rising to 25 per hour. Maximum in 2016 is just before Full Moon, so conditions are not particularly favourable. The parent body is Comet 8P/Tuttle.

The constellation of Orion dominates the sky during this period of the year, and is a useful starting point for recognizing other constellations in the southern sky. Here, orange Betelgeuse, blue-white Rigel and the pinkish Orion Nebula are prominent. Orion can be found in the southern part of the sky (see next page).

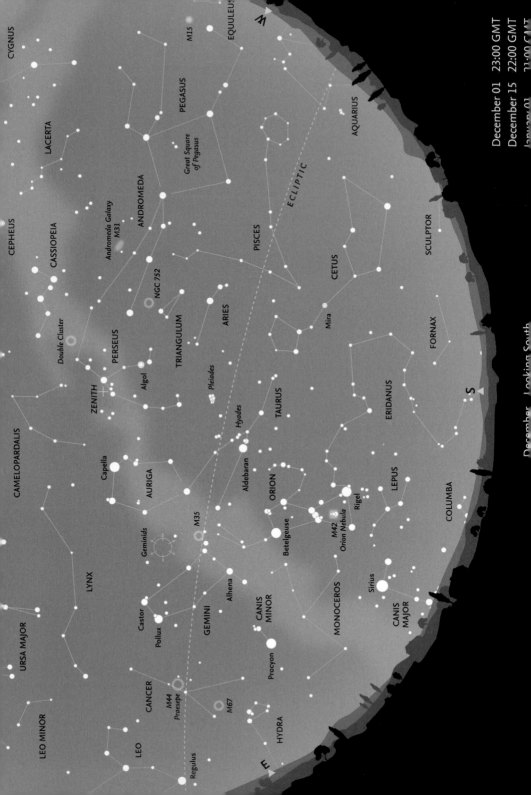

December 01   23:00 GMT
December 15   22:00 GMT
January 01   21:00 GMT

December   Looking South

December – Looking South

The fine open cluster of the *Pleiades* is due south around 22:00, with the *Hyades* cluster, *Aldebaran* and the rest of *Taurus* clearly visible to the east. *Auriga* (with *Capella*) and *Gemini* (with *Castor* and *Pollux*) are both well-placed for observation. *Orion* has made a welcome return to the winter sky, and both *Canis Minor* (with *Procyon*) and *Canis Major* (with *Sirius*, the brightest star in the sky) are now well above the horizon. The small, poorly known constellation of *Lepus* lies to the south of Orion. In the west, *Aquarius* has now disappeared, and *Cetus* is becoming lower, but *Pisces* is still easily seen, as are the constellations of *Aries*, *Triangulum* and *Andromeda* above it. The Great Square of *Pegasus* is starting to plunge down towards the western horizon, and because of its orientation on the sky appears more like a large diamond, standing on one point, than a square.

The constellation of Taurus contains two contrasting open clusters: the compact Pleiades, with its striking blue-white stars, and the more scattered, but much closer, 'V'-shaped Hyades. Orange Aldebaran (α Tauri) is not related to the Hyades, but lies between it and the Earth.

The Moon's phases for December

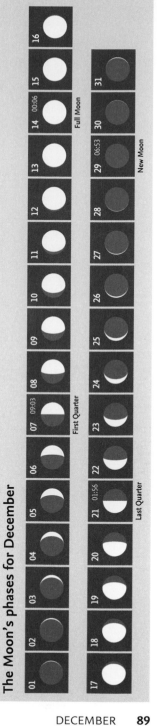

December – Moon and Planets

The Moon

Early in the month, the Moon passes *Mercury*, *Venus* and *Mars* in the evening sky (although Mercury is very low and difficult to see). It passes 5.8° north of Venus on December 2 and 3.0° north of Mars on December 5. On December 13 there is yet another occultation of *Aldebaran*, but only the disappearance is visible from Britain, the Moon setting before reappearance. On December 15–16, just after Full Moon, the Moon passes through the area bounded by *Procyon* (α Canis Minoris), *Alhena* (γ Geminorum) and *Pollux* (β Geminorum). The Moon is close to *Regulus* on December 18. On December 22–23, the Moon is in the same area as *Jupiter* and *Spica* in Virgo. On December 26–27, the Moon is close to *Saturn* (in *Ophiuchus*) and Antares (α Scorpii).

The Planets

Mercury is moving away from the Sun and reaches greatest elongation (20.8°E) on December 11 at magnitude -0.5. It is too low, however to be easily detected. During the month *Venus* is initially in *Sagittarius* in the early evening western sky but brightens from magnitude -4.2 to -4.4 as it moves rapidly through *Capricornus*. It is just inside *Aquarius* on December 31. *Mars* fades from magnitude 0.6 to 0.9 as it moves from *Capricornus* into *Aquarius*. *Jupiter* (magnitude -1.8 to -1.9) is moving slowly eastwards in *Virgo*. *Saturn* (magnitude 0.5) is in *Ophiuchus*, where it has been for the whole year. *Uranus*, at magnitude 5.7–5.8, is retrograding slowly and nearing ζ Piscium. *Neptune* (magnitude 7.9) it also still moving slowly westwards in *Aquarius* but will not reach its stationary point (at which the motion becomes direct) until July 2017.

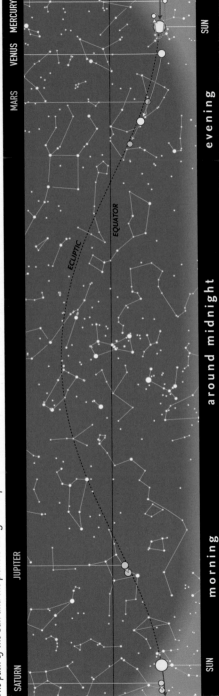

The path of the Sun and the planets along the ecliptic in December.

Calendar for December

Date	Time	Event
02	12:34	Venus 5.8°S of Moon
04–16		Geminid meteor shower
05	10:40	Mars 3.0°S of Moon
07	09:03	First Quarter
11	04:39	Mercury greatest elongation (20.8° E, mag. −0.5)
12	23:29	Moon at perigee (358,461 km)
13–14		Geminid shower maximum
13	04:37	Occultation of Aldebaran
14	00:06	Full Moon
16	05:18	Pollux 10.2°N of Moon
17–23		Ursid meteor shower
18	18:38	Regulus 1.0°N of Moon
21–22		Ursid shower maximum
21	01:56	Last Quarter
21	10:44	Winter solstice
22	16:40	Jupiter 2.4°S of Moon
22	21:59	Spica 6.1°S of Moon
25	05:55	Moon at apogee (405,870 km)
26	20:27	Antares 9.7°S of Moon
27	20:42	Saturn 3.6°S of Moon
28	18:48	Mercury at inferior conjunction
29	06:53	New Moon

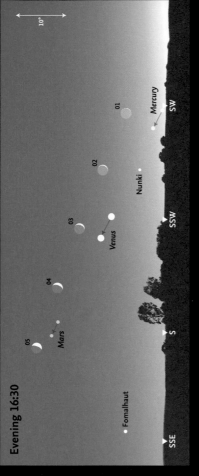

Evening 16:30

December 1–5 • *The crescent Moon passes Mercury, Nunki, Venus and Mars, shortly after sunset. Mercury is very low and difficult to see. Fomalhaut is in the south-southeast.*

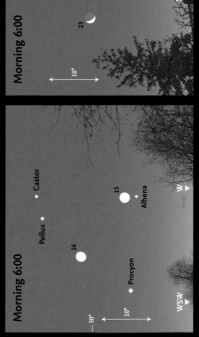

Morning 6:00

December 15–16 • *The Moon with Alhena, Pollux and Procyon, in the morning sky.*

Morning 6:00

December 22–23 • *The Moon passes Spica.*

Glossary and Tables

aphelion	The point on an orbit that is farthest from the Sun.
apogee	The point on its orbit at which the Moon is farthest from the Earth.
appulse	The apparently close approach of two celestial objects; two planets, or a planet and star.
astronomical unit	(AU) The mean distance of the Earth from the Sun, 149,597,870 km.
celestial equator	The great circle on the celestial sphere that is in the same plane as the Earth's equator.
celestial sphere	The apparent sphere surrounding the Earth on which all celestial bodies (stars, planets, etc.) seem to be located.
conjunction	The point in time when two celestial objects have the same celestial longitude. In the case of the Sun and a planet, superior conjunction occurs when the planet lies on the far side of the Sun (as seen from Earth). For Mercury and Venus, inferior conjuction occurs when they pass between the Sun and the Earth.
direct motion	Motion from west to east on the sky.
ecliptic	The apparent path of the Sun across the sky throughout the year. Also: the plane of the Earth's orbit in space.
elongation	The point at which an inferior planet has the greatest angular distance from the Sun, as seen from Earth.
equinox	The two points during the year when night and day have equal duration. Also: the points on the sky at which the ecliptic intersects the celestial equator. The vernal (spring) equinox is of particular importance in astronomy.
gibbous	The stage in the sequence of phases at which the illumination of a body lies between half and full. In the case of the Moon, the term is applied to phases between First Quarter and Full, and between Full and Last Quarter.
inferior planet	Either of the planets Mercury or Venus, which have orbits inside that of the Earth.
magnitude	The brightness of a star, planet or other celestial body. It is a logarithmic scale, where larger numbers indicate fainter brightness. A difference of 5 in magnitude indicates a difference of 100 in actual brightness, thus a first-magnitude star is 100 times as bright as one of sixth magnitude.
meridian	The great circle passing through the North and South Poles of a body and the observer's position; or the corresponding great circle on the celestial sphere that passes through the North and South Celestial Poles and also through the observer's zenith.
nadir	The point on the celestial sphere directly beneath the observer's feet, opposite the zenith.
occultation	The disappearance of one celestial body behind another, such as when stars or planets are hidden behind the Moon.
opposition	The point on a superior planet's orbit at which it is directly opposite the Sun in the sky.
perigee	The point on its orbit at which the Moon is closest to the Earth.
perihelion	The point on an orbit that is closest to the Sun.
retrograde motion	Motion from east to west on the sky.
superior planet	A planet that has an orbit outside that of the Earth.
vernal equinox	The point at which the Sun, in its apparent motion along the ecliptic, crosses the celestial equator from south to north. Also known as the First Point of Aries.
zenith	The point directly above the observer's head.
zodiac	A band, 8° on either side of the ecliptic, within which the Moon and planets appear to move. It consists of twelve equal areas, originally named after the constellation that once lay within it.

The Greek Alphabet

α	Alpha	ε	Epsilon	ι	Iota	ν	Nu	ρ	Rho	φ (φ)	Phi
β	Beta	ζ	Zeta	κ	Kappa	ξ	Xi	σ (ς)	Sigma	χ	Chi
γ	Gamma	η	Eta	λ	Lambda	o	Omicron	τ	Tau	ψ	Psi
δ	Delta	θ (ϑ)	Theta	μ	Mu	π	Pi	υ	Upsilon	ω	Omega

The Constellations

There are 88 constellations covering the whole of the celestial sphere, but 24 of these in the southern hemisphere can never be seen (even in part) from the latitude of Britain and Ireland, so are omitted from this table. The names themselves are expressed in Latin, and the names of stars are frequently given by Greek letters followed by the genitive of the constellation name. The genitives and English names of the various constellations are included.

Name	Genitive	Abbr.	English name
Andromeda	Andromeda	And	Andromeda
Antlia	Antliae	Ant	Air Pump
Aquarius	Aquarii	Aqr	Water Bearer
Aquila	Aquilae	Aql	Eagle
Aries	Arietis	Ari	Ram
Auriga	Aurigae	Aur	Charioteer
Boötes	Boötis	Boo	Herdsman
Camelopardalis	Camelopardalis	Cam	Giraffe
Cancer	Cancri	Cnc	Crab
Canes Venatici	Canum Venaticorum	CVn	Hunting Dogs
Canis Major	Canis Majoris	CMa	Big Dog
Canis Minor	Canis Minoris	CMi	Little Dog
Capricornus	Capricorni	Cap	Sea Goat
Cassiopeia	Cassiopeiae	Cas	Cassiopeia
Centaurus	Centauri	Cen	Centaur
Cepheus	Cephei	Cep	Cepheus
Cetus	Ceti	Cet	Whale
Columba	Columbae	Col	Dove
Coma Berenices	Coma Berenicis	Com	Berenice's Hair
Corona Australis	Coronae Australis	CrA	Southern Crown
Corona Borealis	Coronae Borealis	CrB	Northern Crown
Corvus	Corvi	Crv	Crow
Crater	Crateris	Crt	Cup
Cygnus	Cygni	Cyg	Swan
Delphinus	Delphini	Del	Dolphin
Draco	Draconis	Dra	Dragon
Equuleus	Equulei	Equ	Little Horse
Eridanus	Eridani	Eri	River Eridanus
Fornax	Fornacis	For	Furnace
Gemini	Geminorum	Gem	Twins
Hercules	Herculis	Her	Hercules
Hydra	Hydrae	Hya	Water Snake

Name	Genitive	Abbr.	English name
Lacerta	Lacertae	Lac	Lizard
Leo	Leonis	Leo	Lion
Leo Minor	Leonis Minoris	LMi	Little Lion
Lepus	Leporis	Lep	Hare
Libra	Librae	Lib	Scales
Lupus	Lupi	Lup	Wolf
Lynx	Lyncis	Lyn	Lynx
Lyra	Lyrae	Lyr	Lyre
Microscopium	Microscopii	Mic	Microscope
Monoceros	Monocerotis	Mon	Unicorn
Ophiuchus	Ophiuchi	Oph	Serpent Bearer
Orion	Orionis	Ori	Orion
Pegasus	Pegasi	Peg	Pegasus
Perseus	Persei	Per	Perseus
Pisces	Piscium	Psc	Fishes
Piscis Austrinus	Piscis Austrini	PsA	Southern Fish
Puppis	Puppis	Pup	Stern
Pyxis	Pyxidis	Pyx	Compass
Sagitta	Sagittae	Sge	Arrow
Sagittarius	Sagittarii	Sgr	Archer
Scorpius	Scorpii	Sco	Scorpion
Sculptor	Sculptoris	Scl	Sculptor
Scutum	Scuti	Sct	Shield
Serpens	Serpentis	Ser	Serpent
Sextans	Sextantis	Sex	Sextant
Taurus	Tauri	Tau	Bull
Triangulum	Trianguli	Tri	Triangle
Ursa Major	Ursae Majoris	UMa	Great Bear
Ursa Minor	Ursae Minoris	UMi	Lesser Bear
Vela	Velorum	Vel	Sails
Virgo	Virginis	Vir	Virgin
Vulpecula	Vulpeculae	Vul	Fox

Some common asterisms

Belt of Orion	δ, ε, and ζ Orionis
Big Dipper	α, β, γ, δ, ε, ζ, and η Ursae Majoris
Circlet	γ, θ, ι, λ, and κ Piscium
Guards (or Guardians)	β and γ Ursae Minoris
Head of Cetus	α, γ, ξ^2, μ, and λ Ceti
Head of Draco	β, γ, ξ, and ν Draconis
Head of Hydra	δ, ε, ζ, η, ρ, and σ Hydrae
Keystone	ε, ζ, η, and π Herculis
Kids	ε, ζ, and η Aurigae
Little Dipper	β, γ, η, ζ, ε, δ, and α Ursae Minoris
Lozenge	= Head of Draco
Milk Dipper	ζ, γ, σ, ϕ, and λ Sagittarii
Plough	α, β, γ, δ, ε, ζ, and η Ursae Majoris
Pointers	α and β Ursae Majoris
Sickle	α, η, γ, ζ, μ, and ε Leonis
Square of Pegasus	α, β, and γ Pegasi with α Andromedae
Sword of Orion	θ and ι Orionis
Teapot	γ, ε, δ, λ, ϕ, σ, τ, and ζ Sagittarii
Wain (or Charles' Wain)	= Plough
Water Jar	γ, η, κ, and ζ Aquarii
Y of Aquarius	= Water Jar

Acknowledgements

peresanz/Shutterstock: p.87 (Orion)

Steve Edberg, La Cañada, California: all other constellation photographs

Adam Evans, CC by 2.0: p.21 (Andromeda Galaxy)

ESA: p.63 (Map of Comet 67P)

NASA: p.75 (Map of Ceres)

NASA/ESA/STScI: p.57 (Pluto and satellites)

Ken Sperber, California: p.69 (Double Cluster)

Editorial support was provided by Radmila Topalovic, Astronomy Programmes Officer at the Royal Observatory Greenwich, part of Royal Museums Greenwich.

Further Information

Books

Bone, Neil (1993), *Observer's Handbook: Meteors*, George Philip, London & Sky Publ. Corp., Cambridge, Mass.

Cook, J., ed. (1999), *The Hatfield Photographic Lunar Atlas*, Springer-Verlag, New York

Dunlop, Storm (1999), *Wild Guide to the Night Sky*, HarperCollins, London

Dunlop, Storm (2012), *Practical Astronomy*, 3rd edn, Philip's, London

Dunlop, Storm, Rükl, Antonin & Tirion, Wil (2005), *Collins Atlas of the Night Sky*, HarperCollins, London

Ridpath, Ian, ed. (1998), *Norton's Star Atlas*, 19th edn, Longman, London

Ridpath, Ian, ed. (2003), *Oxford Dictionary of Astronomy*, 2nd edn, Oxford University Press, Oxford

Ridpath, Ian & Tirion, Wil (2004), *Collins Gem - Stars*, HarperCollins, London

Ridpath, Ian & Tirion, Wil (2011), *Collins Pocket Guide Stars and Planets*, 4th edn, HarperCollins, London

Ridpath, Ian & Tirion, Wil (2012), *Monthly Sky Guide*, 9th edn, Cambridge University Press

Rükl, Antonín (1990), *Hamlyn Atlas of the Moon*, Hamlyn, London & Astro Media Inc., Milwaukee

Rükl, Antonín (2004), *Atlas of the Moon*, Sky Publishing Corp., Cambridge, Mass.

Scagell, Robin (2000), *Philip's Stargazing with a Telescope*, George Philip, London

Tirion, Wil (2011), *Cambridge Star Atlas*, 4th edn, Cambridge University Press, Cambridge

Tirion, Wil & Sinnott, Roger (1999), *Sky Atlas 2000.0*, 2nd edn, Sky Publishing Corp., Cambridge, Mass. & Cambridge University Press, Cambridge

Journals

Astronomy, Astro Media Corp., 21027 Crossroads Circle, P.O. Box 1612, Waukesha, WI 53187-1612 USA. http://www.astronomy.com

Astronomy Now, Pole Star Publications, PO Box 175, Tonbridge, Kent TN10 4QX UK. http://www.astronomynow.com

Sky at Night Magazine, BBC publications, London. http://skyatnightmagazine.com

Sky & Telescope, Sky Publishing Corp., Cambridge, MA 02138-1200, USA. http://www.skyandtelescope.com/

Societies

British Astronomical Association, Burlington House, Piccadilly, London W1J 0DU. http://www.britastro.org/
The principal British organization for amateur astronomers (with some professional members), particularly for those interested in carrying out observational programmes. Its membership is, however, worldwide. It publishes fully refereed, scientific papers and other material in its well-regarded Journal.

Federation of Astronomical Societies, Secretary: Ken Sheldon, Whitehaven, Maytree Road, Lower Moor, Pershore, Worcs. WR10 2NY. http://www.fedastro.org.uk/fas/
An organization that is able to provide contact information for local astronomical societies in the United Kingdom.

Royal Astronomical Society, Burlington House, Piccadilly, London W1J 0BQ. http://www.ras.org.uk/
The premier astronomical society, with membership primarily drawn from professionals and experienced amateurs. It has an exceptional library and is a designated centre for the retention of certain classes of astronomical data. Its publications are the standard medium for dissemination of astronomical research.

Society for Popular Astronomy, 36 Fairway, Keyworth, Nottingham NG12 5DU.
>   http://www.popastro.com/
>   A society for astronomical beginners of all ages, which concentrates on increasing members' understanding and enjoyment, but which does have some observational programmes. Its journal is entitled *Popular Astronomy*.

Software

Planetary, Stellar and Lunar Visibility, (Planetary and eclipse freeware): Alcyone Software, Germany.
>   http://www.alcyone.de

Redshift, Redshift-Live. http://www.redshift-live.com/en/

Starry Night & Starry Night Pro, Sienna Software Inc., Toronto, Canada. http://www.starrynight.com

Internet sources

There are numerous sites with information about all aspects of astronomy, and all of those have numerous links. Although many amateur sites are excellent, treat any statements and data with caution. The sites listed below offer accurate information. Please note that the URLs may change. If so, use a good search engine, such as Google, to locate the information source.

Information

Astronomical data (inc. eclipses) HM Nautical Almanac Office: http://astro/ukho/gov.uk

Auroral information Michigan Tech: http://www.geo.mtu.edu/weather/aurora/

Comets JPL Solar System Dynamics: http://ssd.jpl.nasa.gov/

American Meteor Society: http://amsmeteors.org/

Deep-sky objects Saguaro Astronomy Club Database: http://www.virtualcolony.com/sac/

Eclipses NASA Eclipse Page: http://eclipse.gsfc.nasa.gov/eclipse.html

Moon (inc. Atlas) Inconstant Moon: http://www.inconstantmoon.com/

Planets Planetary Fact Sheets: http://nssdc.gsfc.nasa.gov/planetary/planetfact.html

Satellites (inc. International Space Station)
>   Heavens Above: http://www.heavens-above.com/
>   Visual Satellite Observer: http://www.satobs.org/

Star Chart National Geographic Chart:
>   http://www.nationalgeographic.com/features/97/stars/chart/index.html

What's Visible Skyhound: http://www.skyhound.com/sh/skyhound.html
>   Skyview Cafe: http://www.skyviewcafe.com
>   Stargazer: http://www.outerbody.com/stargazer/ (choose 'File, New View', click & drag to change)

Institutes and Organizations

European Space Agency: http://www.esa.int/

International Dark-Sky Association: http://www.darksky.org/

Jet Propulsion Laboratory: http://www.jpl.nasa.gov/

Lunar and Planetary Institute: http://www.lpi.usra.edu/

National Aeronautics and Space Administration: http://www.hq.nasa.gov/

Solar Data Analysis Center: http://umbra.gsfc.nasa.gov/

Space Telescope Science Institute: http://www.stsci.edu/public.html